Two Phase Sampling

Two Phase Sampling

By

Zahoor Ahmad, Muhammad Qaiser Shahbaz
and Muhammad Hanif

CAMBRIDGE
SCHOLARS
PUBLISHING

Two Phase Sampling,
by Zahoor Ahmad, Muhammad Qaiser Shahbaz and Muhammad Hanif

This book first published 2013

Cambridge Scholars Publishing

12 Back Chapman Street, Newcastle upon Tyne, NE6 2XX, UK

British Library Cataloguing in Publication Data
A catalogue record for this book is available from the British Library

ISBN (10): 1-4438-4595-7, ISBN (13): 978-1-4438-4595-3

CONTENTS

Section-IV: Single and Two Phase Sampling with Auxiliary Attributes

PREFACE

This book attempts to give an up to date account and the developments of two-phase sampling using simple random sampling without replacement at each phase. This book also provides a comprehensive review of the previous developments of various researchers in single and two-phase sampling, along with recent developments of the present authors. This book may be used as reference and as a textbook for undergraduate and post-graduate students. This book consists of 9 chapters divided into four sections.

Section one deals with single phase sampling for single and two auxiliary variables, while section two deals with two-phase sampling using single and two auxiliary variables. In section three we have given estimators which are based on multi-auxiliary variables, and section four deals with estimators based on attributes.

We are thankful to Dr. Ken Brewer, Dr. Munir Ahmad, and Dr. A. Z Memon for the clarification of some concepts. We are indebted to Dr. Inam-ul-Haq, Dr Hammad Naqvi, Ms Munaza Bajwa, Ms Asma Tahir and Ms Saria Bano for checking the derivations of mean square errors, some estimators, and for conduct of empirical studies. In general we are thankful to all the M.Phil. and Ph.D. students of National College of Business and Economics, Lahore, for their help at various stages.

—Zahoor Ahmad, M. Qaiser Shahbaz and Muhammad Hanif

CHAPTER ONE

BASIC SAMPLING THEORY

1.1 Introduction

Most survey work involves sampling from finite populations. There are two parts of any sampling strategy. First there is the selection procedure, the manner in which sampling units are to be selected from a finite population. Second there is an estimation procedure, which prescribes how inferences are to be made from sample to population. These inferences may be either enumerative or analytical. Enumerative inference seeks only to describe the finite population under study, whereas analytical inference attempts to explain the underlying distributional and functional characteristics of a population. Enumerative inference typically concerns the estimation of some parameters of a population such as means, totals, proportion and ratios. Viewing the same population analytically we might be interested to regress household income on such variables like number of employed adults, educational level of household head etc. The fitted regression model is an explanatory model. Analytical inference consists of appropriate specification of a model which adequately describes the sample, and hence the population from which it was selected. It is customary to distinguish between enumerative and analytical inference in terms of complexity of the population characteristics of estimator. Estimating a mean, total, etc., is regarded as an enumerative problem, whereas estimating a regression or correlation coefficient is seen as an analytical one, but if the mean in question is a parameter of a simple explanatory model, the inference is enumerative.

Analytical and enumerative inferences used by certain models provide their own probability structure, which guides directions for further inferences. The same methodology is used in other areas of statistics. For enumerative inference, a quite different probability structure is used; it depends on the manner in which the sample is selected. This is the classical finite population sampling inference developed by Neyman (1934), who based his results on Gram (1883) and Bowley (1913) [Brewer and Hanif (1983)].

The sampling theory attracted the most attention as compared with other fields of statistics in the post-war period. This was probably due to methods used in sampling which have many practical applications. It is sufficient to recall that the use of sampling theory has entirely changed the work of data collection and seems to be the reason why most of the names among contemporary statisticians have devoted much of their time to some problems in the theory and practices of sample surveys.

Systematic interest in the use of sampling theory appeared towards the end of the last century when Kiaer (1890) used the representative method for collecting data independently of the census. In 1901 he also demonstrated empirically that stratification could provide good estimates of finite population totals and means. On the recommendation of Kiaer, the International Statistical Institute in 1903 adopted stratified sampling with proportional allocation as an acceptable method of data collection. In line with the present thinking, the selection procedure used by Kiaer and others deals with a method of drawing a representative sample from the population. The primary difference in the earlier use of the sampling methods lay in the selection of the sample. Earlier it was thought that a purposive selection based on the sampler's knowledge of the population with regard to closely correlated characteristic was the best way of getting a sample that could be considered representative of the population. Neyman (1934) published his revolutionary paper in which he made it clear that random selection had its basis on a sound scientific theory, which definitely gave sample survey the character of an objective research tool and made it possible to predict the validity of the survey estimates. Actually in this sense this is marked as the beginning of a new era. Neyman (1934) suggested optimum allocation in stratified sampling. Nowadays, sampling theory has been so developed in its application that it is widely used in all the fields of life and the use of this theory is proved to be fruitful in the underdeveloped areas as well.

1.2 The Use of Auxiliary Information

The history of use of auxiliary information in survey sampling is as old as the history of survey sampling. Graunt (1662) seems to be the first statistical scientist who estimated the population of England based on classical ratio estimate using auxiliary information. At some time in the mid-1780s (the exact date is difficult to establish), the eminent mathematician Laplace (1783) started to press the ailing French government to conduct an enumeration of the population in about 700 communes scattered over the Kingdom (Bru, 1988), with a view to estimating the total population of

France. He intended to use for this purpose the fact that there was already a substantially complete registration of births in all communes, of which there would then have been of the order of 10,000. He reasoned that if he also knew the populations of those sample communes, he could estimate the ratio of population to annual births, and apply that ratio to the known number of births in a given year, to arrive at what we would now describe as a ratio estimate of the total French population (Laplace, 1783, 1814a and 1814b). For various reasons, however, notably the ever-expanding borders of the French empire during Napoleon's early years, events militated against him obtaining a suitable total of births for the entire French population, so his estimated ratio was never used for its original purpose (Bru, 1988; Cochran, 1977, Hald, 1998; Laplace, 1814a and 1814b, p. 762). He did, however, devised an ingenious way for estimating the precision with which that ratio was measured. This was less straightforward than the manner in which it would be estimated today but, at the time, it was a very considerable contribution to the theory of survey sampling.

The works of Bowley (1926) and Neyman (1934, 1938) are the foundation stones of modern sampling theory. These two authors presented use of stratified random sampling and they also put forward theoretical criticism on non-random sampling (purposive sampling). These two classical works may perhaps be referred to as the initial work in the history of survey sampling where use of auxiliary information has been demonstrated. It was a general intuitive of the survey statisticians during the 1930s that the customary method of estimating the population mean or total of a variable of interest (estimand), say y, may be improved to give higher precision of estimation if the information supplied by a related variable (auxiliary variable, supplementary variable, concomitant variable, ancillary variable or benchmark variable), say x, is incorporated intelligibly in the estimation procedure. The work of Watson (1937) and Cochran (1940, 1942) was also the initial work making use of auxiliary information in devising estimation procedures leading to improvement in precision of estimation. Hansen and Hurwitz (1943) were the first to suggest the use of auxiliary information in selecting the population with varying probabilities. In most of the survey situations the auxiliary information is always available. It may either be readily available or may be made available without much difficulty by diverting a part of the survey resources. The customary sources of obtaining relevant auxiliary information on one or more variables are various: i.e. census data, previous surveys or pilot surveys. The auxiliary information may in fact be available in various forms. Some of them may be enumerated as follows.

For details see Tripathi (1970) and Das (1988). In brief, four ways are given as:

i) The values of one or more auxiliary variables may be available a priori only for some units of a finite population,
ii) Values of one or more parameters of auxiliary variables, i.e. population mean(s), population proportion(s), variance, coefficient(s) of variation, Coefficient of skewness may be known. In other words one or more parameters are known,
iii) The exact values of the parameters are not known but their estimated values are known,

And

iv) The values of one or more auxiliary variables may be known for all units of a finite population.

It is worth mentioning here that in addition to the information on auxiliary variable x (in one form or another), information about the estimated variable y may also be available in summary form in some cases. For example coefficient of variation C_y of y may be known exactly [Searle (1964)], or approximate mean square error may be available.

Various research works during the last 70 years confirmed the view that whatever the form of availability of auxiliary information, one may always utilize it to devise sampling strategies which are better (if not uniformly then at least in a part of parametric space) than those in which no auxiliary information is utilized. The method of utilization of auxiliary information depends on the form in which it is available.

As mentioned by Tripathi (1970, 1973, 1976), the auxiliary information in survey sampling may be utilized in the following four ways:

i) The auxiliary information may be used at the planning or designing stage of a survey, i.e. in stratifying the population, the strata can be made according to auxiliary information.
ii) The auxiliary information may be used for the purpose of estimation, i.e. ratio, regression, difference and product estimators etc.
iii) The auxiliary information may be used while selecting the sample, i.e. probabilities proportional to size sampling.

iv) The auxiliary information may be used in mixed ways, i.e. combining at least two of the above.

The classical ratio and regression estimators [Cochran (1940, 42)], difference estimator [Hansen et al. (1953)] and product estimator [Robson (1957), Murthy (1964)] for population mean of estimand variable, which are based on the knowledge of population mean $\bar{X}$ of an auxiliary variable *X,* are well known in the sample survey literature and their detailed study in case of simple random sampling without replacement, and stratified random sampling are available in various books, i.e. [Yates (1960), Deming (1960), Kish (1965), Murthy (1967), Raj (1968), Cochran (1977), and Sukhatme et al. (1984), etc.]. Later these classical estimators were extended to various sampling designs by several researchers. The ratio and regression estimators being biased, the efforts were made by various researchers to obtain unbiased ratio and regression type estimators during the 1950s and 1960s. Hartly and Ross (1954) defined an unbiased ratio estimator based on simple random sampling using the knowledge of auxiliary variable. Basically four methods have been used to obtain unbiased or almost unbiased estimators.

i) Bias correction using the knowledge of $\bar{X}$ [Hartley and Ross (1954), Beale (1962) and Tin (1965)]
ii) Jack-knife technique [Quenouille (1956), Durbin (1959) and Mickey (1959).
iii) Mahalanobis's technique of interpenetrating sub-samples. [Murthy and Nanjamma (1959)]
iv) Varying probabilities technique based on the information of X. [Lahiri (1951), Sen (1952), Midzuno (1952) and Ikeda (1952) reported by Midzuno (1952)].

Similar methods were used to obtain an unbiased regression estimator [Williams (1963), Robson (1957)]. Srivenkataramana (1980) and Srivenkataramana and Tracy (1979, 80, 81) were the first to introduce transformation of auxiliary variable X to change negative correlation situation into a positive one and vice-versa, giving duals to ratio estimator. Tripathi and Singh (1988) gave a generalized transformation capable of dealing with positive and negative correlation situations simultaneously and derived a wide class of unbiased product type estimators which also includes the unbiased product type estimators due to T.J. Rao (1983,87), Singh et al. (1987) and several others.

The study of ratio and regression estimators based on super-population models have been made by several authors during the 1960s and 1970s. In this context we refer to Brewer (1963), P. S. R. S. Rao (1969), Royall (1970), Cassel et al. (1977), etc. The performance of ratio estimator based on small samples has been studied by P. S. R. S. Rao and J. N. K. Rao (1971) and others.

The use of multivariate auxiliary information in defining ratio, difference, regression, product, and ratio-cum-product type estimators etc. for population mean $\bar{Y}$, under various situations, has been considered by Olkin (1958), Raj (1965), Srivastava (1965, 1966, 71), Shukla (1965, 1966), Rao and Mudholkar (1967), Singh (1967), Khan and Tripathi (1967), Tripathi (1970, 1976), etc. Tripathi (1987) unified most of the results by considering a class of estimators based on general sampling design and multivariate information.

The problem of estimation of population ratio $R = \frac{Y}{X}$ or $R = \frac{Y}{Z}$ using auxiliary variable did not attract much attention during the 1950s. The work relating to the use of auxiliary information in estimating α was initiated by J. N. K. Rao (1957), Singh (1965, 1967, 1969), Rao and Pereira (1968), and Tripathi (1980). This work was followed by several authors. The use of auxiliary information in the form of coefficient of variation C_x of an auxiliary variable X for estimation of Y was considered first by Das and Tripathi (1980). Das and Tripathi (1978) also initiated the work related to the use of auxiliary information for estimating the finite population variance σ_y^2. Das and Tripathi (1979) considered the use of σ_x^2. Srivastava and Jhajj (1981) and Das and Tripathi (1981) considered the simultaneous utilization of $\bar{X}$ and σ_x^2 for estimation of $\bar{Y}$. Srivatava and Jhajj (1980) also considered the use of $\bar{X}$ and σ_x^2 for estimating σ_y^2. The use of more than one auxiliary variable in defining the selection probabilities to select the population units with varying probabilities and with replacement was considered by Maiti and Tripathi (1976), and Agarwal and Singh (1980).

1.3 Difference between Multi-stage and Multi-phase Sampling

The differences between multistage and multiphase sampling are given below:

- The first difference is that in multistage sampling the natures (and most specifically the sizes) of the population units differ from stage to stage; whereas in multiphase sampling there is only one kind of population unit, and each such unit could end up being selected, either just at the first phase or at one or more consecutive subsequent phases also.
- The second difference, which builds on the first, is that the units at any stage are larger than units from the stage immediately following. (They must be larger, being actually comprised of them! and consequently, they themselves also make up units from the immediately previous stage.)

For example, in an Australian Labour Force Survey, the first stage units were typically Local Government Areas, containing thousands of dwellings; the second stage units were typically Census Collectors' Districts containing typically two or three hundred dwellings each; the third stage units were typically street blocks, each usually containing a few dozen dwellings; and the fourth stage units were typically the individual dwellings themselves (though occasionally they were houses containing more than one household; and therefore, by definition, more than one dwelling). By contrast, in multi-phase sampling, any unit of the population can become a first phase sample unit, and, if selected at the first stage, can also be a second phase sample unit, and so on.

- The third difference is that in multistage sampling, information about the actual survey variables is gathered only from the units selected at the final stage, e.g. from all the people living in a selected dwelling, but not from the remainder of the people in a selected Block (or Collector's District or Local Government Area). By contrast in multiphase sampling, we typically collect only a little information from (or, more often, a little information about) first phase sample units. It has to be relatively cheap information, available perhaps from administrative records; otherwise there would be no point in using a multiphase sample at all. It also has to be information that is correlated with (in practice usually more or less proportional to) the survey variables of interest.
- A fourth commonly occurring difference, though not an essential one, is that multistage sampling can involve three or sometimes even four stages, though it is quite uncommon to have more than two phases. For this reason, another expression for two-phase sampling is "double sampling".

- The last difference we can think of is that one usually has to work one's way up the stages by calculating (say for a three-stage sample) first, third stage sample estimates for the relevant two-stage sample units; next, use those estimates to arrive at first stage sample estimates, and finally use those estimates in turn to estimate figures for the entire population. In multiphase sampling, only the entire population is estimated. It can be viewed as being estimated (rather poorly) on the basis of the lowest phase sample only, rather better by incorporating information from the next phase up, and so on, with the most accurate estimate being made on the basis of information from all the phases of selection together. Each phase of the sample, other than the last, would therefore be supplying only an adjustment factor of the order of unity.

1.4 Notations

Consider a population of N units. Associated with the Ith unit are the variables of main interest Y_I and the auxiliary variables X_I and Z_I for $I = 1, 2, 3, \ldots\ldots, N$. Let $\bar{X}$, $\bar{Z}$ and $\bar{Y}$ be the population mean of variables *X*, *Z* and *Y* respectively. Further let S_x^2, S_z^2 and S_y^2 be corresponding variances. The coefficient of variation of these variables will be denoted by $C_x^2 = \left(S_x^2/\bar{X}^2\right)$, $C_y^2 = \left(S_y^2/\bar{Y}^2\right)$ and $C_z^2 = \left(S_z^2/\bar{Z}^2\right)$ respectively. Also ρ_{xy}, ρ_{xz} and ρ_{yz} will represent population correlation coefficients between *X* and *Y*, *X* and *Z*, and *Y* and *Z* respectively.

The first phase sample means based on n_1 units are denoted by $\bar{y}_1$, $\bar{x}_1$ and $\bar{z}_1$ etc. and second phase sample means are denoted by $\bar{y}_2, \bar{x}_2$ and $\bar{z}_2$. We will also use $\theta_1 = n_1^{-1} - N^{-1}$ and $\theta_2 = n_2^{-1} - N^{-1}$ so that $\theta_2 > \theta_1$. For notational purpose we will assume that the mean of estimand (estimated variable) and auxiliary variables can be approximated from their population means so that $\bar{x}_h = \bar{X} + \bar{e}_{x_h}$, where $\bar{x}_h$ is sample mean of auxiliary variable *X* at *h*–th phase for *h* = 1 and 2. Similar notation will be used for other quantitative auxiliary variables. For qualitative auxiliary variables we will use $p_h = P + \bar{e}_{\tau_h}$. In using the relation between sample and population mean of auxiliary variable we will assume that $\left|\bar{e}_{x_h}\right|$ is much smaller as compared with $\left|\bar{X}\right|$. For variable of interest we will use $\bar{y}_h = \bar{Y} + \bar{e}_{y_h}$ with usual assumptions.

In this monograph we will use following expectations for deriving the mean square error of estimators which are based upon quantitative auxiliary variables:

$$\left.\begin{aligned}
&E\left(\bar{e}_{x_1}^2\right)=\theta_1\bar{X}^2C_x^2\ ;\ E\left(\bar{e}_{z_1}^2\right)=\theta_1\bar{Z}^2C_z^2\ ;\ E\left(\bar{e}_{y_1}^2\right)=\theta_1\bar{Y}^2C_y^2\\
&E\left(\bar{e}_{x_2}^2\right)=\theta_2\bar{X}^2C_x^2\ ;\ E\left(\bar{e}_{z_2}^2\right)=\theta_2\bar{Z}^2C_z^2\ ;\ E\left(\bar{e}_{y_2}^2\right)=\theta_2\bar{Y}^2C_y^2\\
&E\left(\bar{e}_{x_1}\bar{e}_{x_2}\right)=\theta_1\bar{X}^2C_x^2\ ;\ E\left(\bar{e}_{z_1}\bar{e}_{z_2}\right)=\theta_1\bar{Z}^2C_z^2\\
&E\left(\bar{e}_{x_1}\bar{e}_{y_1}\right)=\theta_1\bar{X}\bar{Y}C_xC_y\rho_{xy}\ ;\ E\left(\bar{e}_{x_1}\bar{e}_{y_2}\right)=E\left(\bar{e}_{x_2}\bar{e}_{y_1}\right)=\theta_1\bar{X}\bar{Y}C_xC_y\rho_{xy}\\
&E\left(\bar{e}_{x_2}\bar{e}_{y_2}\right)=\theta_2\bar{X}\bar{Y}C_xC_y\rho_{xy}\ ;\ E\left(\bar{e}_{x_1}\bar{e}_{z_1}\right)=\theta_1\bar{X}\bar{Z}C_xC_z\rho_{xz}\ \ etc\\
&E\left(\bar{e}_{x_1}-\bar{e}_{x_2}\right)^2=\left(\theta_2-\theta_1\right)\bar{X}^2C_x^2\ \ etc\\
&E\left\{\bar{e}_{y_1}\left(\bar{e}_{x_1}-\bar{e}_{x_2}\right)\right\}=E\left\{\bar{e}_{z_1}\left(\bar{e}_{x_1}-\bar{e}_{x_2}\right)\right\}=0\\
&E\left\{\bar{e}_{y_2}\left(\bar{e}_{x_1}-\bar{e}_{x_2}\right)\right\}=\left(\theta_1-\theta_2\right)\bar{X}\bar{Y}C_xC_y\rho_{xy}\ \ etc
\end{aligned}\right\}$$

(1.4.1)

In case of several auxiliary variables; say q; the sample mean of ith auxiliary at hth phase will be denoted by $\bar{x}_{(i)h} = \bar{X}_i + \bar{e}_{x_{(i)h}}$. The expectations in (1.4.1) are modified as:

$$\left.\begin{aligned}
&E\left(\bar{e}^2_{x_{(i)1}}\right)=\theta_1\bar{X}_i^2C_{x_i}^2\ ;\ E\left(\bar{e}^2_{x_{(i)2}}\right)=\theta_2\bar{X}_i^2C_{x_i}^2\\
&E\left(\bar{e}_{x_{(i)1}}\bar{e}_{x_{(i)2}}\right)=\theta_1\bar{X}_i^2C_{x_i}^2\ ;\ E\left(\bar{e}_{x_{(i)1}}\bar{e}_{x_{(k)2}}\right)=\theta_1\bar{X}_i\bar{X}_kC_{x_i}C_{x_k}\rho_{ik}\ ;\\
&E\left(\bar{e}_{x_{(i)1}}\bar{e}_{y_1}\right)=\theta_1\bar{X}_i\bar{Y}C_{x_i}C_y\rho_{yx_i}\ ;\ E\left(\bar{e}_{x_{(i)2}}\bar{e}_{y_2}\right)=\theta_2\bar{X}_i\bar{Y}C_{x_i}C_y\rho_{yx_i}\\
&E\left(\bar{e}_{x_{(i)1}}-\bar{e}_{x_{(i)2}}\right)^2=\left(\theta_2-\theta_1\right)\bar{X}_i^2C_{x_i}^2\ \ etc\\
&E\left\{\bar{e}_{y_2}\left(\bar{e}_{x_{(i)1}}-\bar{e}_{x_{(i)2}}\right)\right\}=\left(\theta_1-\theta_2\right)\bar{X}_i\bar{Y}C_{x_i}C_y\rho_{yx_i}\ \ etc
\end{aligned}\right\}$$

(1.4.2)

The collection of correlation coefficients of various variables will be embedded in the following matrices:

$\mathbf{R}_{y\mathbf{x}}$: Correlation matrix for all variables under study

$\mathbf{R}_{\mathbf{x}}$: Correlation matrix for all auxiliary variables

$\mathbf{R}_{yx_i}$: Minor of $\mathbf{R}_{y\mathbf{x}}$ corresponding to ρ_{yx_i}

The following popular result of Arora and Lal (1989) will also be used:

$$\left|\mathrm{R}_{yx}\right|/\left|\mathrm{R}_x\right|=\left(1-\rho^2_{y.\mathrm{x}}\right). \tag{1.4.3}$$

We will also use the vector notations in case of multiple auxiliary variables and the sample mean vector of auxiliary variables at h^{th} phase will be denoted by $\bar{\mathrm{x}}_h$ with relation $\bar{\mathrm{x}}_h=\bar{\mathrm{X}}+\bar{\mathrm{e}}_{x_h}$. The following additional expectations are also useful:

$$E\left(\bar{\mathrm{e}}_{x_h}\bar{\mathrm{e}}'_{x_h}\right)=\theta_h\mathrm{S}_x\ ,\ E\left(\bar{e}_{y_1}\bar{\mathrm{e}}_{x_h}\right)=\theta_1\mathrm{s}_{yx}\ ;h\ge 1 \tag{1.4.4}$$

Similar expectations for qualitative auxiliary variables are:

$$\left.\begin{aligned}
&E\left(\bar{e}_{\tau_1}^2\right)=\theta_1 P_\tau^2 C_\tau^2 \ ;\ E\left(\bar{e}_{\delta_1}^2\right)=\theta_1 P_\delta^2 C_\delta^2 \ ;\ E\left(\bar{e}_{y_1}^2\right)=\theta_1 \bar{Y}^2 C_y^2\\
&E\left(\bar{e}_{\tau_2}^2\right)=\theta_2 P_\tau^2 C_\tau^2 \ ;\ E\left(\bar{e}_{\delta_2}^2\right)=\theta_2 P_\delta^2 C_\delta^2 \ ;\ E\left(\bar{e}_{y_2}^2\right)=\theta_2 \bar{Y}^2 C_y^2\\
&E\left(\bar{e}_{\tau_1}\bar{e}_{\tau_2}\right)=\theta_1 P_\tau^2 C_\tau^2 \ ;\ E\left(\bar{e}_{\delta_1}\bar{e}_{\delta_2}\right)=\theta_1 P_\delta^2 C_\delta^2\\
&E\left(\bar{e}_{\tau_1}\bar{e}_{y_1}\right)=\theta_1 \bar{Y}P_\tau C_\tau C_y \rho_{pb} ; E\left(\bar{e}_{\tau_1}\bar{e}_{y_2}\right)=E\left(\bar{e}_{\tau_2}\bar{e}_{y_1}\right)=\theta_1 \bar{Y}P_\tau C_\tau C_y \rho_{pb}\\
&E\left(\bar{e}_{\tau_2}\bar{e}_{y_2}\right)=\theta_2 \bar{Y}P_\tau C_\tau C_y \rho_{pb} \ ;\ E\left(\bar{e}_{\tau_1}\bar{e}_{\delta_1}\right)=\theta_1 P_\tau P_\delta C_\tau C_\delta Q_{xz} \ etc\\
&E\left(\bar{e}_{\tau_1}-\bar{e}_{\tau_2}\right)^2=\left(\theta_2-\theta_1\right)P_\tau^2 C_\tau^2 \ etc\\
&E\left\{\bar{e}_{y_2}\left(\bar{e}_{\tau_1}-\bar{e}_{\tau_2}\right)\right\}=\left(\theta_1-\theta_2\right)\bar{Y}P_\tau C_\tau C_y \rho_{pb} \ etc
\end{aligned}\right\} \quad (1.4.5)$$

For multivariate case we will use following notations:

$$\left.\begin{aligned}
&E\left(\bar{\mathrm{e}}_{\mathrm{y}_1}\bar{\mathrm{e}}_{\mathrm{y}_1}^{/}\right)=\theta_1 \mathrm{S}_\mathrm{y} \ ;\ E\left(\bar{\mathrm{e}}_{\mathrm{x}_1}\bar{\mathrm{e}}_{\mathrm{x}_1}^{/}\right)=\theta_1 \mathrm{S}_\mathrm{x}\\
&E\left(\bar{\mathrm{e}}_{\mathrm{y}_2}\bar{\mathrm{e}}_{\mathrm{y}_2}^{/}\right)=\theta_2 \mathrm{S}_\mathrm{y} \ ;\ E\left(\bar{\mathrm{e}}_{\mathrm{x}_2}\bar{\mathrm{e}}_{\mathrm{x}_2}^{/}\right)=\theta_2 \mathrm{S}_\mathrm{x}\\
&E\left(\bar{\mathrm{e}}_{\mathrm{y}_1}\bar{\mathrm{e}}_{\mathrm{x}_1}^{/}\right)=\theta_1 \mathrm{S}_\mathrm{yx} \ ;\ E\left(\bar{\mathrm{e}}_{\mathrm{x}_1}\bar{\mathrm{e}}_{\mathrm{y}_1}^{/}\right)=\theta_1 \mathrm{S}_\mathrm{xy}\\
&E\left(\bar{\mathrm{e}}_{\mathrm{y}_2}\bar{\mathrm{e}}_{\mathrm{x}_2}^{/}\right)=\theta_2 \mathrm{S}_\mathrm{yx} \ ;\ E\left(\bar{\mathrm{e}}_{\mathrm{x}_2}\bar{\mathrm{e}}_{\mathrm{y}_2}^{/}\right)=\theta_2 \mathrm{S}_\mathrm{xy}\\
&E\left(\bar{\mathrm{e}}_{\mathrm{y}_1}\bar{\mathrm{e}}_{\mathrm{x}_2}^{/}\right)=\theta_1 \mathrm{S}_\mathrm{yx}=E\left(\bar{\mathrm{e}}_{\mathrm{y}_2}\bar{\mathrm{e}}_{\mathrm{x}_1}^{/}\right)\\
&E\left(\bar{\mathrm{e}}_{\mathrm{x}_1}\bar{\mathrm{e}}_{\mathrm{y}_2}^{/}\right)=\theta_1 \mathrm{S}_\mathrm{xy}=E\left(\bar{\mathrm{e}}_{\mathrm{x}_2}\bar{\mathrm{e}}_{\mathrm{y}_1}^{/}\right)
\end{aligned}\right\} \quad (1.4.6)$$

where $\mathbf{S}_\mathbf{y}$ is covariance matrix of variables of interest and so on. We will use the notation $t_{1(1)}, t_{2(1)}, t_{3(1)},.....$ to denote single-phase sampling estimators whereas estimators for two phase sampling will be denoted by $t_{1(2)}, t_{2(2)}, t_{3(2)},.....$ etc. We will also use the notations $n, \theta, \bar{x}, \bar{y}$ *and* $\bar{z}$ etc. for sample size and mean(s) whenever only single-phase sampling is involved. The notations $n_1, \theta_1, \bar{x}_1, \bar{y}_1$ *and* $\bar{z}_1$ will not be used in that case. We will also use $\bar{x} = \bar{X} + \bar{e}_x$ etc. to describe the relationship between sample mean and population mean of a variable whenever only single-phase sampling is involved.

1.5 Mean per Unit Estimator in Simple Random Sampling

The mean per unit estimator is perhaps the oldest estimator in the history of survey sampling. This estimator has provided the basis for the development of a number of single phase and two phase sampling estimators. The estimator for a sample of size n drawn from a population of size N is defined as:

$$t_0 = \frac{1}{n}\sum_{i=1}^{n} y_i = \bar{y}. \tag{1.5.1}$$

Using the notation defined in Section (1.4), i.e. $\bar{y} = \bar{Y} + \bar{e}_y$, we can write $t_0 - \bar{Y} = \bar{e}_y$. The mean square error (variance; as estimator is unbiased) can be immediately written as:

$$MSE(t_0) = \theta \bar{Y}^2 C_y^2. \tag{1.5.2}$$

The mean square error of t_0; given in (1.5.2) has been the basis in numerous comparative studies. The expression (1.5.2) has also been used by a number of survey statisticians to see the design effect of newly developed sampling designs.

Searle (1964) presented a modified version of mean per unit estimator as given below:

Suppose

$$t_s = k\bar{y}, \tag{1.5.3}$$

where k is a constant which is determined by minimizing mean square error of (1.5.3). The mean square error of (1.5.3) is readily written as:

$$MSE(t_s) = E\left(t_s - \bar{Y}\right)^2 = E\left(k\,\bar{y} - \bar{Y}\right)^2. \tag{1.5.4}$$

Expanding the square and rearranging the terms, the mean square error of t_s may be written as:

$$MSE(t_s) = E\left[k^2\left(\bar{y} - \bar{Y}\right)^2 + (k-1)^2\bar{Y}^2 + 2k(k-1)\bar{Y}\left(\bar{y} - \bar{Y}\right)\right]$$

Applying expectation and simplifying we may write the mean square error as:

$$MSE(t_s) \approx \bar{Y}^2\left[\theta k^2 C_y^2 + (k-1)^2\right]. \tag{1.5.5}$$

Differentiating (1.5.5) *w.r.t.* k and equating to zero, we obtain optimum value of k which minimizes (1.5.4) as:

$$k = \left(1 + \theta C_y^2\right)^{-1}. \tag{1.5.6}$$

Using k from (1.5.6) in (1.5.5) and simplifying, the mean square error of t_s is:

$$MSE\left(t_s\right) \approx \frac{\theta \bar{Y}^2 C_y^2}{1 + \theta C_y^2} = \frac{MSE\left(t_0\right)}{1 + \bar{Y}^{-2} MSE\left(t_0\right)} \tag{1.5.7}$$

Comparison of (1.5.7) with (1.5.2) shows that the estimator proposed by Searle (1964) is always more precise than the mean per unit estimator.

The estimator proposed by Searle (1964) is popularly known as Shrinkage mean per unit estimator. Many statisticians have studied the shrinkage version of popular survey sampling estimators. The focus of these studies had been reduction in mean square error of estimate or increase in precision. We will present some popular shrinkage estimators in this monograph. A general system of development of shrinkage estimators has been proposed by Shahbaz and Hanif (2009).

1.6 Shrinkage Estimator by Shahbaz and Hanif

Suppose a population parameter Θ can be estimated by using an estimator $\hat{\eta}$ with mean square error $MSE\left(\hat{\eta}\right)$. Shahbaz and Hanif (2009) defined a general shrinkage estimator as $\hat{\eta}_s = d\hat{\eta}$ where d is a constant to be determined so that mean square error of $\hat{\eta}_s$ is minimized. The optimum value of d and mean square error of $\hat{\eta}_s$ can be derived by assuming that $\hat{\eta} = \Theta + \bar{e}_\eta$ with $E\left(\bar{e}_\eta\right) = 0$ and $E\left(\bar{e}_\eta^2\right) = MSE\left(\hat{\eta}\right)$. The mean square error of $\hat{\eta}_s$ and optimum value of d; under the above assumption is derived below:

The mean square error of $\hat{\eta}_s$ is given as:

$$MSE\left(\hat{\eta}_s\right) = E\left(\hat{\eta}_s - \Theta\right)^2 = E\left(d\hat{\eta} - \Theta\right)^2 = E\left[d\left(\Theta + \bar{e}_\eta\right) - \Theta\right]^2. \tag{1.6.1}$$

Differentiating (1.6.1) *w.r.t.* d and equating to zero we have:

$$E\left[\left\{d\left(\Theta + \bar{e}_\eta\right) - \Theta\right\}\left(\Theta + \bar{e}_\eta\right)\right] = 0, \tag{1.6.2}$$

or $d\left[\Theta^2+E\left(\bar{e}_\eta^2\right)+2\Theta E\left(\bar{e}_\eta\right)\right]-\Theta^2-\Theta E\left(\bar{e}_\eta\right)=0$.

Now using the fact that $E\left(\bar{e}_\eta\right)=0$ and $E\left(\bar{e}_\eta^2\right)=MSE\left(\hat{\eta}\right)$, the above equation may be written as:

$$d\left[\Theta^2+MSE\left(\hat{\eta}\right)\right]-\Theta^2=0,$$

or $d=\Theta^2\left[\Theta^2+MSE\left(\hat{\eta}\right)\right]^{-1}=\left[1+\Theta^{-2}MSE\left(\hat{\eta}\right)\right]^{-1}$. (1.6.3)

Now from (1.6.1) we may write:

$$MSE\left(\hat{\eta}_s\right)=E\left[\left\{d\left(\Theta+\bar{e}_\eta\right)-\Theta\right\}\left\{d\left(\Theta+\bar{e}_\eta\right)-\Theta\right\}\right],$$

or $MSE\left(\hat{\eta}_s\right)=E\left[\left\{d\left(\Theta+\bar{e}_\eta\right)-\Theta\right\}d\left(\Theta+\bar{e}_\eta\right)\right]-\Theta E\left[\left\{d\left(\Theta+\bar{e}_\eta\right)-\Theta\right\}\right]$.
(1.6.4)

Also from (1.6.2) we have $E\left[\left\{d\left(\Theta+\bar{e}_\eta\right)-\Theta\right\}d\left(\Theta+\bar{e}_\eta\right)\right]=0$. Then (1.6.4) will be

$$MSE\left(\hat{\eta}_s\right)=-\Theta E\left[\left\{d\left(\Theta+\bar{e}_\eta\right)-\Theta\right\}\right]=-\Theta\left[dE\left(\bar{e}_\eta\right)-\Theta\left(1-d\right)\right],$$

or $MSE\left(\hat{\eta}_s\right)=\Theta^2\left(1-d\right)$ (1.6.4)

Using (1.6.3) in (1.6.4), the mean square error of $\hat{\eta}_s$ is:

$$MSE\left(\hat{\eta}_s\right)=\frac{MSE\left(\hat{\eta}\right)}{1+\Theta^{-2}MSE\left(\hat{\eta}\right)}. \qquad (1.6.5)$$

The expression for mean square error given in (1.6.5) may be used to obtain the mean square error of the shrinkage version of any estimator. Also the shrinkage version of any estimator can be obtained by multiplying (1.6.3) with estimator to be shrunk. The estimator proposed by Searle (1964) turned out to be a special case of the shrinkage estimator proposed by Shahbaz and Hanif (2009) by using $\hat{\eta}=\bar{y}$.

An alternative derivation of the above estimator suggested by H.P. Singh (2011) is given as:

Consider

$$\hat{\eta}_s=d\hat{\eta}, \qquad (1.6.6)$$

or $\left(\hat{\eta}_s-\Theta\right)=\left(d\hat{\eta}-\Theta\right)=\left\{d\left(\hat{\eta}-\Theta\right)+\left(d-1\right)\Theta\right\}$. (1.6.7)

Squaring both sides of (1.6.7) we have:

$$\left(\hat{\eta}_s-\Theta\right)^2=\left\{d\left(\hat{\eta}-\Theta\right)+\left(d-1\right)\Theta\right\}^2,$$
$$=d^2\left(\hat{\eta}-\Theta\right)^2+\left(d-1\right)^2\Theta^2+2d\left(d-1\right)\Theta\left(\hat{\eta}-\Theta\right).$$

Taking expectation of both sides of the above expression we get the MSE of $\hat{\eta}_s$ as

$$MSE\left(\hat{\eta}_s\right)=d^2MSE\left(\hat{\eta}\right)+\Theta^2\left(d-1\right)^2+2d\left(d-1\right)\Theta B\left(\hat{\eta}\right)$$
$$=\left[d^2\left\{\Theta^2+2\Theta B\left(\hat{\eta}\right)+MSE\left(\hat{\eta}\right)\right\}-2d\left\{\Theta^2+\Theta B\left(\hat{\eta}\right)\right\}+\Theta^2\right]. \tag{1.6.8}$$

Expression (1.6.8) is minimized when

$$d=d_{opt}=\frac{\left\{\Theta^2+\Theta B\left(\hat{\eta}\right))\right\}}{\left\{\Theta^2+2\Theta B\left(\hat{\eta}\right)+MSE\left(\hat{\eta}\right)\right\}}. \tag{1.6.9}$$

Thus the resulting minimum MSE of $\hat{\eta}_s$ is given by

$$MSE_{\min}\left(\hat{\eta}_s\right)=\left[\Theta^2-\frac{\left\{\Theta^2+\Theta B\left(\hat{\eta}\right)\right\}^2}{\left\{\Theta^2+2\Theta B\left(\hat{\eta}\right)+MSE\left(\hat{\eta}\right)\right\}}\right],$$
$$=\frac{\Theta^2\left[MSE\left(\hat{\eta}\right)-\left\{B\left(\hat{\eta}\right)\right\}^2\right]}{\left[\Theta^2+2\Theta B\left(\hat{\eta}\right)+MSE\left(\hat{\eta}\right)\right]}. \tag{1.6.10}$$

If we assume that $\hat{\eta}$ is an unbiased estimator (i.e. $Var\left(\hat{\eta}\right)=MSE\left(\hat{\eta}\right)$ then (1.6.10) reduces to

$$MSE_{\min}\left(\hat{\eta}_s\right)=\frac{\Theta^2MSE\left(\hat{\eta}\right)}{\Theta^2+MSE\left(\hat{\eta}\right)}=\frac{MSE(\hat{\eta})}{1+\Theta^{-2}MSE\left(\hat{\eta}\right)}, \tag{1.6.11}$$

which is identical to (1.6.5) suggested by Shahbaz and Hanif (2009).

SECTION-I:

SINGLE-PHASE SAMPLING WITH AUXILIARY VARIABLE

CHAPTER TWO

SINGLE-PHASE SAMPLING USING ONE AND TWO AUXILIARY VARIABLES

2.1 Introduction

Survey statisticians are always trying to estimate population characteristics with greater precision. The precision of estimate can be increased by using two methodologies. Firstly the precision may be increased by using adequate sampling design for the estimated variable (estimand). Secondly the precision may be increased by using an appropriate estimation procedure, i.e. some auxiliary information which is closely associated with the variable under study.

The ratio and regression methods of estimation are two classical methods of estimation in survey sampling which use information about auxiliary variables. These two methods of estimation are more efficient than the mean per unit method of estimation in terms of smaller mean square error. Various estimators have been proposed in the literature that use information on multiple auxiliary variables and consequently produce smaller error as compared with estimators based upon single auxiliary variables.

Various authors have proposed mixed type estimators, i.e. use of both ratio and regression estimators in some fashion. These mixed estimators have been seen performing better as compared with individual estimators. Mohanty (1967) used this methodology for the first time to propose mixed estimator using two auxiliary variables.

In this Chapter we will explore some popular single-phase sampling estimators which utilize information of one and two auxiliary variables. We will first reproduce some estimators which are based upon single auxiliary variable and then we will describe those estimators which are based upon two auxiliary variables.

2.2 Estimators using Single Auxiliary Variable

Suppose that information on a single auxiliary X is available for complete population units from some previous source. This available information of X may be very useful in increasing the precision of estimate. In the following we describe some popular estimators which are based upon information of a single auxiliary variable.

2.2.1 Classical Ratio Estimator

The classical ratio estimator is perhaps the most celebrated estimator in single phase sampling. The estimator based upon single auxiliary variable is given as:

$$t_{1(1)} = \bar{y}\frac{\bar{X}}{\bar{x}} = \frac{\bar{y}}{\bar{x}}\bar{X}. \tag{2.2.1}$$

Using the notations from section (1.4) we may write the estimator (2.2.1) as:

$$t_{1(1)} = \left(\bar{Y}+\bar{e}_y\right)\frac{\bar{X}}{\bar{X}+\bar{e}_x} = \left(\bar{Y}+\bar{e}_y\right)\left(1+\frac{\bar{e}_x}{\bar{X}}\right)^{-1}.$$

Expanding $\left(1+\bar{e}_x/\bar{X}\right)^{-1}$, under the assumption that $\left|\bar{e}_x/\bar{X}\right|<1$, and retaining the first term only and by using argument of Cochran (1940); we have:

$$t_{1(1)} = \left(\bar{Y}+\bar{e}_y\right)\left(1-\frac{\bar{e}_x}{\bar{X}}\right) = \bar{Y}+\bar{e}_y-\frac{\bar{Y}}{\bar{X}}\bar{e}_x.$$

The mean square error of $t_{1(1)}$ is

$$\text{MSE}\left(t_{1(1)}\right) = E\left(t_{1(1)}-\bar{Y}\right)^2 = E\left(\bar{e}_y-\frac{\bar{Y}}{\bar{X}}\bar{e}_x\right)^2. \tag{2.2.2}$$

Squaring, applying expectation and simplifying we get

$$MSE\left(t_{1(1)}\right) = \theta\bar{Y}^2\left(C_y^2-2C_xC_y\rho_{xy}+C_x^2\right),$$

or $$MSE\left(t_{1(1)}\right) = \theta\bar{Y}^2\left[C_y^2-2C_xC_y\rho_{xy}+C_x^2+C_y^2\rho_{xy}^2-C_y^2\rho_{xy}^2\right],$$

or $$MSE\left(t_{1(1)}\right) = \theta\bar{Y}^2\left[C_y^2-2C_xC_y\rho_{xy}+C_x^2+C_y^2\rho_{xy}^2-C_y^2\rho_{xy}^2\right]. \tag{2.2.3}$$

Comparing (2.2.3) with variance of mean per unit estimator we can see that the ratio estimator will have larger precision as compared with mean per unit estimator if

$$\rho_{xy} > C_x/2C_y \cdot \tag{2.2.4}$$

For a smaller correlation coefficient, the mean per unit estimator will be more precise. Survey statisticians have proposed some modifications of classical ratio estimator from time to time. In the following we discuss some of the popular modifications of the classical ratio estimator.

2.2.2 Prasad's (1989) Ratio Estimator

Prasad (1989) proposed following modification of ratio estimator following the lines of Searle (1964):

$$t_{2(1)} = \frac{k\bar{y}}{\bar{x}}\bar{X}, \tag{2.2.5}$$

where k is constant. Using Section (2.5), $\beta_i = -\left(1-2k_i\right)\bar{Y}/\bar{X}_i$ may be written as

$$t_{2(1)} = k\bar{Y}\left(1+\frac{\bar{e}_y}{\bar{Y}}\right)\left(1+\frac{\bar{e}_x}{\bar{X}}\right)^{-1},$$

or $$t_{2(1)} = k\bar{Y}\left[1+\frac{\bar{e}_y}{\bar{Y}}-\frac{\bar{e}_x}{\bar{X}}-\frac{\bar{e}_y\bar{e}_x}{\bar{X}\bar{Y}}+\frac{\bar{e}_x^2}{\bar{X}^2}\right],$$

or $$t_{2(1)}-\bar{Y} = \bar{Y}\left[k\left(1+\frac{\bar{e}_y}{\bar{Y}}-\frac{\bar{e}_x}{\bar{X}}-\frac{\bar{e}_y\bar{e}_x}{\bar{X}\bar{Y}}+\frac{\bar{e}_x^2}{\bar{X}^2}\right)-1\right].$$

Squaring the above equation and ignoring higher order terms we have:

$$\left(t_{2(1)}-\bar{Y}\right)^2 = \bar{Y}^2\left[k^2\left\{1+\frac{\bar{e}_y^2}{\bar{Y}^2}+\frac{3\bar{e}_x^2}{\bar{X}^2}+\frac{2\bar{e}_y}{\bar{Y}}-\frac{2\bar{e}_x}{\bar{X}}-\frac{4\bar{e}_x\bar{e}_y}{\bar{Y}\bar{X}}\right\}\right.$$
$$\left.+1-2k\left\{1+\frac{\bar{e}_y}{\bar{Y}}-\frac{\bar{e}_x}{\bar{X}}-\frac{\bar{e}_y\bar{e}_x}{\bar{X}\bar{Y}}+\frac{\bar{e}_x^2}{\bar{X}^2}\right\}\right].$$

Applying expectations on the above equation, the mean square error of $t_{2(1)}$, to first degree of approximation is

$$MSE\left(t_{2(1)}\right) = \bar{Y}^2\left[k^2\left\{1+\theta(C_y^2+3C_x^2-4\rho_{xy}C_xC_y)\right\}\right.$$

$$+1-2k\left\{1+\theta(C_x^2-\rho_{xy}C_xC_y)\right\}\Big], \qquad (2.2.6)$$

which is minimum when

$$k=k_{opt}=\frac{\left\{1+\theta\left(C_x^2-\rho_{xy}C_xC_y\right)\right\}}{\left\{1+\theta\left(C_y^2+3C_x^2-4\rho_{xy}C_xC_y\right)\right\}}.$$

Using k_{opt} in (2.2.6) and simplifying, the mean square of $t_{2(1)}$ is:

$$MSE_{\min}\left(t_{2(1)}\right)=\bar{Y}^2\left[1-\frac{\left\{1+\theta\left(C_x^2-\rho_{xy}C_xC_y\right)\right\}^2}{\left\{1+\theta\left(C_y^2+3C_x^2-4\rho_{xy}C_xC_y\right)\right\}}\right],$$

or $$MSE\left(t_{2(1)}\right)=\frac{\left[MSE\left(t_{1(1)}\right)-\left\{B\left(t_{1(1))}\right)\right\}^2\right]}{\left[1+RMSE\left(t_{1(1)}\right)+2RB\left(t_{1(1)}\right)\right]}, \qquad (2.2.7)$$

where $RB\left(t_{1(1)}\right)=\bar{Y}^{-1}B\left(t_{1(1)}\right)$ and $RMSE\left(t_{1(1)}\right)=\bar{Y}^{-2}MSE\left(t_{1(1)}\right)$.

The above result can be easily obtained from the general result given in Section 1.6.

2.2.3 Product Estimator

The ratio estimator and its versions, which we will discuss later, are useful for estimating population mean and total when the auxiliary variable has a high positive correlation with the variable of interest. Sometimes the available auxiliary variable has a negative correlation with the variable of interest; the survey statisticians have proposed alternative methods of estimation in such situations. One such method of estimation is the product method which was initially studied by Robson (1957) and Murthy (1964). Cochran (1977), Chaubey et al. (1984), and Kaur (1984) have also done substantial work on it. The conventional product estimator is

$$t_{3(1)}=\bar{y}\frac{\bar{x}}{\bar{X}}. \qquad (2.2.8)$$

Using (1.4) in (2.2.8) and on simplification we get:

$$t_{3(1)}-\bar{Y}=\bar{e}_y+\frac{\bar{Y}}{\bar{X}}\bar{e}_x .$$

Squaring, applying expectation and using results from section 1.4, the mean square error of (2.2.8) is:

$$MSE\left(t_{3(1)}\right)=\theta\left[\bar{Y}^2C_y^2+2\frac{\bar{Y}}{\bar{X}}\bar{Y}\,\bar{X}C_yC_x\rho_{xy}+\frac{\bar{Y}^2}{\bar{X}^2}\bar{X}^2C_x^2\right].$$

After simplification we get

$$MSE\left(t_{3(1)}\right)=\theta\bar{Y}^2\left[C_y^2+2C_yC_x\rho_{xy}+C_x^2\right],$$

or $$MSE\left(t_{3(1)}\right)=\theta\bar{Y}^2\left[C_y^2(1-\rho_{xy}^2)+(C_x+C_y\rho_{xy})^2\right]. \qquad (2.2.9)$$

Comparing (2.2.9) with variance of mean per unit estimator, we can see that the product estimator will be more efficient than mean per unit estimator if $\rho_{xy}<-C_x/2C_y$ [Murthy, 1964; Kendall and Stuart, 1966; and Robson, 1957]. Some notable versions of product estimator have been proposed by Liu (1990), Upadhyaya et al. (1985), and Singh (1986a, 1986b). We will discuss these generalizations in Chapter 3 of this Monograph.

2.2.4 Classical Regression Estimator

The regression method of estimation is perhaps most popular method of estimation is single and two phase sampling. This method of estimation has been proved to be more efficient when a linear relationship exists between the variable of study and auxiliary variable and the regression line does not passes through origin. The classical regression estimator has been the field of study for many survey statisticians. Some notable references include Srivastava (1967), Walsh (1970), Reddy (1973, 74), Gupta (1978), Sahai (1979), Vos (1980), Adhvaryan and Gupta (1983) and others. The classical regression estimator is given as:

$$t_{4(1)}=\bar{y}+\beta_{yx}\left(\bar{X}-\bar{x}\right). \qquad (2.2.10)$$

The mean square error of (2.2.10) is obtained by using the notations from section (1.4) and writing (2.2.10) as:

$$t_{4(1)}=\bar{Y}+\bar{e}_y+\beta_{yx}\left(\bar{X}-\bar{X}-\bar{e}_x\right),$$

On simplification we get

$$t_{4(1)} - \overline{Y} = \overline{e}_y - \beta_{yx}\overline{e}_x .$$

Squaring and applying expectation, the mean square error is:

$$MSE\left(t_{4(1)}\right) = E\left(\overline{e}_y^2 + \beta_{yx}^2\overline{e}_x^2 - 2\beta_{yx}\overline{e}_y\overline{e}_x\right),$$

or $MSE\left(t_{4(1)}\right) = \theta\left[\overline{Y}^2C_y^2 + \beta_{yx}^2C_x^2\overline{X}^2 - 2\beta_{yx}\overline{YX}\,C_xC_y\rho_{xy}\right].$ (2.2.11)

The optimum value of β_{yx} which minimizes (2.2.11) is $\beta_{yx} = \left(\overline{Y}C_y/\overline{X}C_x\right)\rho_{yx}$. Using this optimum value in (2.2.11) and simplifying, we have following mean square error of (2.2.10):

$$MSE\left(t_{4(1)}\right) \approx \theta\overline{Y}^2C_y^2\left(1-\rho_{xy}^2\right). \tag{2.2.12}$$

Comparing (2.2.12) with variance of mean per unit estimator we can conclude that the classical regression estimator is always more efficient as compared with the mean per unit estimator.

2.2.5 Other Related Estimators

Ratio, product and regression estimators are corner stones in estimation of population characteristics using information on auxiliary variables. The mean square error of regression estimator, given in (2.2.12), shows that the regression estimator will have a smaller mean square error as compared with the mean per unit estimator, whatever correlation exists between the main variable of study and auxiliary variable. Several other estimators have been proposed in the literature which use different mechanisms of estimation. We now describe these estimators which are in equivalence class with the regression estimator.

Rao (1968) proposed the following estimator, which has also been studied by Vos (1980) and Agarwal and Kumar (1980). The estimator is:

$$t_{5(1)} = \overline{y} - W\frac{\overline{y}}{\overline{X}}\left(\overline{x} - \overline{X}\right). \tag{2.2.13}$$

The estimator is identical to mean per unit estimator for $W = 0$. Singh (1969) suggested the following estimator

$$t_{6(1)} = \frac{\overline{y}}{\overline{x} - (1-W)\left(\overline{x} - \overline{X}\right)}\overline{X} . \tag{2.2.14}$$

For $W = 1$ the estimator (2.2.14) is identical to ratio estimator and for $W = 0$ it is equal to the mean per unit estimator.

A mixture of ratio and regression estimator proposed by Walsh (1970); also studied by Reddy (1973, 1974); is given as:

$$t_{7(1)} = \bar{y}\left[\frac{\bar{X}}{\bar{X} + b\left(\bar{x} - \bar{X}\right)}\right], \tag{2.2.15}$$

where b is constant and to be determined such that mean square error of (2.2.15) is minimized. This estimator is like a ratio estimator where ratio has been taken by using a functional form for mean of an auxiliary variable.

Srivenkataramana and Srinath (1976) considered the following estimator of population mean using single auxiliary variable:

$$t_{8(1)} = \bar{y}\left[1 - \frac{1}{1-f}\left(\frac{\bar{x} - \bar{X}}{\bar{x}}\right)\right]\frac{\bar{x}}{\bar{X}},$$

(2.2.16)

where f is a constant to be determined. Srivenkataramana and Tracy (1979) simplified (2.2.16) to propose following estimators of population mean:

$$t_{9(1)} = \bar{y}\left[1 - \frac{f}{\bar{X}}\left(\bar{x} - \bar{X}\right)\right], \tag{2.2.17}$$

$$t_{10(1)} = \bar{y}\left[1 + \frac{f}{\bar{X}}\left(\bar{x} - \bar{X}\right)\right], \tag{2.2.18}$$

$$t_{11(1)} = \bar{y}\left[1 + \frac{1-f}{\bar{X}}\left(\bar{x} - \bar{X}\right)\right], \tag{2.2.19}$$

$$t_{12(1)} = \bar{y}\left[1 - \frac{1}{W - \bar{X}}\left(\bar{x} - \bar{X}\right)\right], \tag{2.2.20}$$

and

$$t_{13(1)} = \bar{y} + \frac{1}{W - \bar{X}}\left(\bar{x} - \bar{X}\right). \text{ (Shukla and Pandey (1982))} \tag{2.2.21}$$

Gupta (1978) further modified Rao's (1968) estimator and proposed the following estimator of population mean:

$$t_{14(1)} = \left\{\bar{y} - \left(1 - W\right)\frac{\bar{y}}{\bar{X}}\left(\bar{x} - \bar{X}\right)\right\}\frac{\bar{X}}{\bar{x}}. \tag{2.2.22}$$

Gupta (1978) also proposed the following estimator of population mean:

$$t_{15(1)} = \bar{y}\left[1 - \frac{(1-W)}{\bar{X}}\left(\bar{x} - \bar{X}\right)\right]\frac{\bar{x}}{\bar{X}}. \tag{2.2.23}$$

This modified estimator is actually a modification of product estimator. Gupta (1978) used the same argument in proposing (2.2.23) which he used to propose (2.2.22). The difference between (2.2.22) and (2.2.23) is that the former estimator modifies the ratio estimator and latter modifies the product estimator.

The difference estimator, suggested by Hansen et al. (1942), is like a regression estimator and is given as:

$$t_{16(1)} = \bar{y} - \lambda\left(\bar{x} - \bar{X}\right). \tag{2.2.24}$$

Another version of difference estimator proposed by Tripathi (1970) independently is:

$$t_{17(1)} = \bar{y} - p\left(\bar{x} - \bar{X}\right). \tag{2.2.25}$$

In fact there is no difference between (2.2.24) and (2.2.25).

Sahai (1979), using Rao's (1968) estimator, proposed following estimator of population mean:

$$t_{18(1)} = \frac{\bar{y}\left[1 + \frac{1-W}{\bar{X}}\left(\bar{x} - \bar{X}\right)\right]}{\bar{x} - (1-W)\left(\bar{x} - \bar{X}\right)}\bar{X}. \tag{2.2.26}$$

The optimum value of W which minimizes mean square error of (2.2.26) is given as $W = \left(1 + \rho_{xy}C_y/C_x\right)/2$. For $W = 1$, the estimator (2.2.26) reduces to classical ratio estimator.

Sahai (1979) proposed the following estimator for population mean in single-phase sampling using a mixture of sample and population mean of auxiliary variable in numerator and denominator.

$$t_{19(1)} = \bar{y}\left[\frac{\alpha_1\bar{X} + (1-\alpha_1)\bar{x}}{\alpha_1\bar{x} + (1-\alpha_1)\bar{X}}\right], \tag{2.2.27}$$

where α_1 is a constant that minimizes mean square error of (2.2.27).

Vos (1980) used weighted average of mean per unit and ratio estimator to propose the following estimator.

$$t_{20(1)} = \alpha\bar{y} + (1-\alpha)\frac{\bar{y}}{\bar{x}}\bar{X}, \tag{2.2.28}$$

where α is constant. The estimator has been independently studied by Advaryu and Gupta (1983). The estimator (2.2.28) is identical to mean per unit estimator for $\alpha = 1$ and is identical to ratio estimator for $\alpha = 0$.

Ray and Sahai (1979) proposed two estimators for populations mean in single phase sampling. The proposed estimators are:

$$t_{21(1)} = \bar{y} + (1-W)\frac{\bar{Y}}{\bar{X}}(\bar{x} - \bar{X}); \tag{2.2.29}$$

and $$t_{22(1)} = \bar{y}\left[1 - \frac{1}{2}\left(1 - \left(\frac{\bar{x}}{\bar{X}}\right)^2\right)\right]\left(\frac{\bar{X}}{\bar{x}}\right)^{1+W}. \tag{2.2.30}$$

Both estimators proposed by Ray and Sahai (1979) are relatively difficult to apply in practical situations.

Tripathi (1980) proposed another estimator using ratio of difference estimators. This estimator is given as:

$$t_{23(1)} = \frac{\bar{y} - p(\bar{x} - \bar{X})}{\bar{x} - q(\bar{x} - \bar{X})}\bar{X}; \tag{2.2.31}$$

Das (1988) proposed the following estimator in single-phase sampling which is a slight modification of classical regression estimator:

$$t_{24(1)} = \bar{y} - W\lambda(\bar{x} - \bar{X}); \tag{2.2.32}$$

For $W = 1$ estimator (2.2.32) reduces to classical regression estimator and for $W = 0$ it reduces to mean per unit estimator.

Some more estimators are available in the literature, i.e.:

$$t_{25(1)} = \bar{y}\left[1 - \frac{\bar{x}^{\beta} - \bar{X}^{\beta}}{\bar{X}^{\beta}}\right], \text{ (Sahai and Ray, 1980)} \tag{2.2.33}$$

$$t_{26(1)} = \bar{y}\left[1 - \frac{W}{\bar{x}^{\beta}}\left(\bar{x}^{\beta} - \bar{X}^{\beta}\right)\right], \text{ (Sisodia and Dwivedi, 1981)} \tag{2.2.34}$$

and

$$t_{27(1)} = \left[\bar{y} \pm \left(\bar{x}^{\frac{1}{2}} - \bar{X}^{\frac{1}{2}}\right)\right]\frac{\bar{X}}{\bar{x}}, \text{ (Singh, 1981)} \tag{2.2.35}$$

$$t_{28(1)} = \bar{y}\left[1 \pm \frac{t_2 - 1}{\bar{x} + (t_1 + t_2 - 1)\bar{X}}(\bar{x} - \bar{X})\right], \tag{2.2.36}$$

(Sahai and Ray, 1980)

$$t_{29(1)} = \bar{y}\left\{k\left(\bar{X}/\bar{x}\right)+\left(1-k\right)\left(\bar{x}/\bar{X}\right)\right\}. \text{ (Singh and Espejo, 2003)} \quad (2.2.37)$$

It is worth mentioning that estimators $t_{5(1)}$ to $t_{29(1)}$ have the same mean square error, which is equal to the mean square error of classical regression estimator to order n^{-1}. These estimators are said to be in equivalence class with the classical regression estimator.

We now discuss some estimators for single-phase sampling which are based upon information of two auxiliary variables in the following section.

2.3 Estimators using Two Auxiliary Variables

We know that the precision of estimates is highly influenced by availability of suitable auxiliary information. This fact has been explored in the previous section where we have discussed estimators based upon a single auxiliary variable. The concept of including auxiliary variables in estimation stages has been extended by increasing the number of variables. The inclusion of a larger number of auxiliary variables at estimation stages will definitely provide more precise estimates. Survey statisticians have extensively explored the role of multiple auxiliary variables. A simple case of inclusion of multiple auxiliary variables is to develop estimators using two auxiliary variables. In the following we will describe some available estimators for single-phase sampling which are based upon two auxiliary variables. The case of inclusion of multiple auxiliary variables will be discussed in later chapters.

2.3.1 Mohanty's (1967) Estimator

Suppose information on two auxiliary variables X and Z is available from a complete population and from sample. Mohanty (1967) suggested an estimator using two auxiliary variables by combining both ratio and regression estimators:

$$t_{30(1)} = \left[\bar{y}+\beta_{yx}\left(\bar{X}-\bar{x}\right)\right]\frac{\bar{Z}}{\bar{z}}. \quad (2.3.1)$$

The estimator (2.3.1) reduces to conventional regression estimator if information on Z is unavailable and it reduces to conventional ratio estimator if information on X is unavailable. The estimator (2.3.1) was termed as regression-in-ratio estimator. For mean square error we use section (1.4) in (2.3.1) and get:

$$t_{30(1)}=\left[\bar{Y}+\bar{e}_y-\beta_{yx}\bar{e}_x\right]\left(1+\frac{\bar{e}_z}{\bar{Z}}\right)^{-1}.$$

Expanding $\left(1+\bar{e}_z/\bar{Z}\right)^{-1}$ and retaining the first term only we have:

$$t_{30(1)}-\bar{Y}=\bar{e}_y-\frac{\bar{Y}}{\bar{Z}}\bar{e}_z-\beta_{yx}\,\bar{e}_x\,.$$

Squaring and applying expectation we have:

$$MSE\left(t_{30(1)}\right)=E\left(t_{30(1)}-\bar{Y}\right)^2=E\left[\bar{e}_y-\frac{\bar{Y}}{\bar{Z}}\bar{e}_z-\beta_{yx}\bar{e}_x\right]^2.$$

Expanding the square and using results from section 1.4 we have:

$$MSE\left(t_{30(1)}\right)=\theta\left[\bar{Y}^2C_y^2+\frac{\bar{Y}^2}{\bar{Z}^2}\bar{Z}^2\,C_z^2+\beta_{yx}^2\,\bar{X}^2\,C_x^2-2\frac{\bar{Y}}{\bar{Z}}\bar{Y}\,\bar{Z}C_yC_z\,\rho_{yz}\right.$$
$$\left.-2\beta_{xy}\bar{X}\bar{Y}C_xC_y\,\rho_{xy}+2\frac{\bar{Y}}{\bar{Z}}\beta_{xy}\bar{X}\,\bar{Z}C_xC_z\rho_{xz}\right]$$

Using $\beta_{yx}=\left(\bar{Y}\,C_y\rho_{xy}\right)/\left(\bar{X}C_x\right)$ in the above equation and simplifying, the mean square error of $t_{30(1)}$ is:

$$MSE\left(t_{30(1)}\right)=\theta\bar{Y}^2\left[C_y^2\left(1-\rho_{xy}^2\right)+\left(C_z-C_y\rho_{yz}\right)^2-C_y^2\,\rho_{yz}^2+2C_y\,C_z\,\rho_{xz}\,\rho_{xy}\right]. \tag{2.3.2}$$

Comparing (2.3.2) with the mean square error of mean per unit estimator, we can see that the Mohanty's (1967) estimator will be more precise if:

$$\left(\rho_{yz}-\rho_{xz}\rho_{xy}\right)\geq\frac{1}{2}\left(\frac{C_z}{C_y}-\frac{C_y}{C_z}\rho_{xy}^2\right). \tag{2.3.3}$$

The estimator (2.3.1) will be more precise than the conventional ratio estimator if:

$$\left(C_x-C_y\rho_{xz}\right)^2\geq\left[\left(C_z-C_y\rho_{yz}\right)^2+C_y\left(2C_z\rho_{xz}\rho_{xy}-C_y\rho_{yz}^2\right)\right]. \tag{2.3.4}$$

Also, the Mohanty's estimator will be more precise than the regression estimator if:

$$\left(\rho_{yz}-\rho_{xy}\rho_{xz}\right)\geq C_z/2C_y\ . \tag{2.3.5}$$

The inequalities given in (2.3.3) – (2.3.5) clearly indicate that Mohanty's (1967) estimator is conditionally more precise than the conventional estimators.

Samiuddin and Hanif (2006) have been very instrumental in developing estimators using information on two auxiliary variables. They have proposed several estimators by augmenting ratio, regression and Mohanty's (1967) estimator. We now discuss these modifications in the following section.

2.4 Modifications by Samiuddin and Hanif (2006)

Various modifications have been proposed by Samiuddin and Hanif (2006). These modifications extended classical ratio and regression estimators alongside the weighted averages of these classical estimation methods. These modifications are given below.

2.4.1 Chain Ratio-Type Estimator by Samiuddin and Hanif (2006)

The classical ratio estimator $t_{1(1)}$ involves one auxiliary variable X. Chand (1975) constructed an estimator in two-phase sampling which uses information on two auxiliary variables. Samiuddin and Hanif (2006) introduced the idea of Chand (1975) in single sampling and proposed the following chain ratio estimator for population mean:

$$t_{30(1)}=\bar{y}\frac{\bar{X}}{\bar{x}}\frac{\bar{Z}}{\bar{z}}\ . \tag{2.4.1}$$

The estimator (2.4.1) is a simple extension of the classical ratio estimator. Using (1.4) in (2.4.1) we get:

$$t_{31(1)}=\left(\bar{Y}+\bar{e}_y\right)\left(1+\bar{e}_x/\bar{X}\right)^{-1}\left(1+\bar{e}_z/\bar{Z}\right)^{-1}.$$

Expanding and retaining the first term only we have:

$$t_{31(1)}=\left(\bar{Y}+\bar{e}_y\right)\left(1-\bar{e}_x/\bar{X}\right)\left(1-\bar{e}_z/\bar{Z}\right)$$

or $t_{31(1)}-\bar{Y}=\bar{e}_y-\dfrac{\bar{Y}}{\bar{X}}\bar{e}_x-\dfrac{\bar{Y}}{\bar{Z}}\bar{e}_z\ .$

The mean square error of $t_{31(1)}$ is:

$$MSE\left(t_{31(1)}\right) \approx E\left(t_{31(1)} - \bar{Y}\right)^2 = E\left(\bar{e}_y - \frac{\bar{Y}}{\bar{X}}\bar{e}_x - \frac{\bar{Y}}{\bar{Z}}\bar{e}_z\right)^2.$$

Squaring, applying expectation and using results from section 1.4, and after simplification the mean square error of (2.4.1) is:

$$MSE\left(t_{31(1)}\right) \approx \theta\bar{Y}^2\left[C_y^2 + C_x^2 + C_z^2 - 2C_xC_y\rho_{xy} - 2C_yC_z\rho_{yz} + 2C_xC_z\rho_{xz}\right]. \quad (2.4.2)$$

Comparing (2.4.2) with (2.2.3), we can see that the chain ratio estimator of Samiuddin and Hanif (2006) is more precise than conventional ratio estimator if:

$$\rho_{xz} > \frac{C_y}{C_x}\rho_{yz}. \quad (2.4.3)$$

From condition (2.4.3) we can see that a strong correlation between auxiliary variables is needed for increased precision of chain ratio estimator. The condition (2.4.3) is sometime very hard to meet in practice.

2.4.2 Revised Ratio Estimator

Samiuddin and Hanif (2006) proposed following revised ratio estimator using two auxiliary variables:

$$t_{32(1)} = a\left(\bar{y}\frac{\bar{X}}{\bar{x}}\right) + (1-a)\left(\bar{y}\frac{\bar{Z}}{\bar{z}}\right). \quad (2.4.4)$$

The estimator (2.4.4) is actually a weighted average of two ratio estimators, one based on auxiliary variable *X* and other based on auxiliary variable *Z*. Using the notations from section 1.4, the estimator (2.4.4) may be written as:

$$t_{32(1)} = a\left(\bar{Y} + \bar{e}_y - \frac{\bar{Y}}{\bar{X}}\bar{e}_x\right) + (1-a)\left[\bar{Y} + \bar{e}_y - \frac{\bar{Y}}{\bar{Z}}\bar{e}_z\right]$$

or $$t_{32(1)} - \bar{Y} = a\left(\bar{e}_y - \frac{\bar{Y}}{\bar{X}}\bar{e}_x\right) + (1-a)\left(\bar{e}_y - \frac{\bar{Y}}{\bar{Z}}\bar{e}_z\right).$$

The mean square error of $t_{32(1)}$ is given as:

$$MSE\left(t_{32(1)}\right) \approx E\left(t_{33(1)} - \bar{Y}\right)^2 = E\left[a\left(\bar{e}_y - \frac{\bar{Y}}{\bar{X}}\bar{e}_x\right) + (1-a)\left(\bar{e}_y - \frac{\bar{Y}}{\bar{Z}}\bar{e}_z\right)\right]^2,$$

or
$$MSE\left(t_{32(1)}\right) \approx E\left[\left(\bar{e}_y - \frac{\bar{Y}}{\bar{Z}}\bar{e}_z\right) - a\left(\frac{\bar{Y}}{\bar{X}}\bar{e}_x - \frac{\bar{Y}}{\bar{Z}}\bar{e}_z\right)\right]^2. \tag{2.4.5}$$

The optimum value of a, which minimizes (2.4.5) is:

$$a = \frac{E\left[\left(\bar{e}_y - \frac{\bar{Y}}{\bar{Z}}\bar{e}_z\right)\left(\frac{\bar{Y}}{\bar{X}}\bar{e}_x - \frac{\bar{Y}}{\bar{Z}}\bar{e}_z\right)\right]}{E\left(\frac{\bar{Y}}{\bar{X}}\bar{e}_x - \frac{\bar{Y}}{\bar{Z}}\bar{e}_z\right)^2}.$$

Using results from section 1.4 in the above equation and simplifying, we get:

$$a = \frac{C_y C_x \rho_{xy} - C_y C_z \rho_{yz} - C_x C_z \rho_{xz} + C_z^2}{C_x^2 + C_z^2 - 2C_x C_z \rho_{xz}}.$$

Using the value of a in (2.4.5) and simplifying the mean square error of $t_{32(1)}$ will be:

$$MSE\left(t_{32(1)}\right) = \theta\bar{Y}^2\left[\left(C_y^2 + C_z^2 - 2\rho_{yz}C_yC_z\right) - \frac{\left(C_xC_y\rho_{xy} - C_yC_z\rho_{yz} - C_xC_z\rho_{xz} + C_z^2\right)^2}{C_x^2 + C_z^2 - 2C_xC_z\rho_{xz}}\right]. \tag{2.4.6}$$

This may alternatively be written as:

$$MSE\left(t_{32(1)}\right) = \theta\bar{Y}^2\left[\left(C_y - C_z^2\rho_{yz}\right)^2 + C_z^2\left(1-\rho_{yz}\right)^2 - \frac{\left(C_xC_y\rho_{xy} - C_yC_z\rho_{yz} - C_xC_z\rho_{xz} + C_z^2\right)^2}{\left(C_x - C_z\rho_{xz}\right)^2 + C_z^2\left(1-\rho_{xz}^2\right)}\right]. \tag{2.4.7}$$

Direct comparison of (2.4.2) and (2.4.7) does not provide clear information about precision of $t_{32(1)}$ as compared with $t_{31(1)}$.

2.4.3 Modification of Mohanty's (1967) Estimator–I

Samiuddin and Hanif (2006) have provided the following modification of Mohanty's (1967) estimator, $t_{30(1)}$:

$$t_{33(1)} = a\left[\bar{y} + b_{yx}\left(\bar{X} - \bar{x}\right)\right]\frac{\bar{Z}}{\bar{z}} + (1-a)\left[\bar{y} + b_{yz}\left(\bar{Z} - \bar{z}\right)\right]\frac{\bar{X}}{\bar{x}} \quad . \quad (2.4.8)$$

We may write (2.4.8) as

$$t_{33(1)} = at_1 + (1-a)t_2,$$

where

$$t_1 = \left[\bar{y} + b_{yx}\left(\bar{X} - \bar{x}\right)\right]\frac{\bar{Z}}{\bar{z}}, \quad (2.4.9)$$

and

$$t_2 = \left[\bar{y} + b_{yz}\left(\bar{Z} - \bar{z}\right)\right]\frac{\bar{X}}{\bar{x}}. \quad (2.4.10)$$

The mean square error of (2.4.8) may be obtained by using the properties of linear combinations as:

$$t_{33(1)} - \bar{Y} = a\left(t_1 - \bar{Y}\right) + (1-a)\left(t_2 - \bar{Y}\right),$$

or $$t_{33(1)} - \bar{Y} = \left(t_2 - \bar{Y}\right) - a\left\{\left(t_2 - \bar{Y}\right) - \left(t_1 - \bar{Y}\right)\right\}.$$

The mean square error of $t_{33(1)}$ is:

$$MSE\left(t_{33(1)}\right) \approx E\left[\left(t_2 - \bar{Y}\right) - a\left\{\left(t_2 - \bar{Y}\right) - \left(t_1 - \bar{Y}\right)\right\}\right]^2. \quad (2.4.11)$$

We first find the optimum value of a, that minimizes mean square error of $t_{32(1)}$ by partially differentiating (2.4.11) *w.r.t.* a and equating to zero. The optimum value is:

$$a = \frac{E\left(t_2 - \bar{Y}\right)^2 - E\left(t_1 - \bar{Y}\right)\left(t_2 - \bar{Y}\right)}{E\left(t_1 - \bar{Y}\right)^2 + E\left(t_2 - \bar{Y}\right)^2 - 2E\left(t_1 - \bar{Y}\right)\left(t_2 - \bar{Y}\right)}. \quad (2.4.12)$$

Further we can write (2.4.9) and (2.4.10) as:

$$t_1 - \bar{Y} = \left(\bar{e}_y - \frac{\bar{Y}}{\bar{Z}} \bar{e}_z - \beta_{yx} \bar{e}_x \right), \tag{2.4.13}$$

and

$$t_2 - \bar{Y} = \left(\bar{e}_y - \frac{\bar{Y}}{\bar{X}} \bar{e}_x - \beta_{yz} \bar{e}_z \right). \tag{2.4.14}$$

Squaring (2.4.13), applying expectations and simplifying we have:

$$E\left(t_1 - \bar{Y}\right)^2 = \theta \bar{Y}^2 \left[C_y^2 \left(1 - \rho_{xy}^2\right) + C_z^2 - 2\rho_{yz} C_y C_Z + 2\rho_{yx} \rho_{xz} C_y C_z \right]. \tag{2.4.15}$$

Similarly squaring (2.4.14), applying expectations and simplifying we have:

$$E\left(t_2 - \bar{Y}\right)^2 = \theta \bar{Y}^2 \left[C_y^2 \left(1 - \rho_{yz}^2\right) + C_x^2 - 2\rho_{yx} C_y C_x + 2\rho_{yz} \rho_{xz} C_y C_x \right]. \tag{2.4.16}$$

Finally, multiplying (2.4.13) and (2.4.14), applying expectations and simplifying we get:

$$E\left(t_1 - \bar{Y}\right)\left(t_2 - \bar{Y}\right) = \theta \bar{Y}^2 \left[C_y^2 \left(1 - \rho_{xy}^2 - \rho_{yz}^2 + \rho_{yx} \rho_{yz} \rho_{xz}\right) + C_x C_z \rho_{xz} \right]. \tag{2.4.17}$$

The optimum value of a can be obtained by using (2.4.15), (2.4.16) and (2.4.17) in (2.4.12). Further, by using (2.4.12) in (2.4.11) and simplifying we may write the mean square error of $t_{33(1)}$ as:

$$MSE\left(t_{33(1)}\right) = E\left(t_2 - \bar{Y}\right)^2 - \frac{\left[E\left(t_2 - \bar{Y}\right)^2 - E\left(t_1 - \bar{Y}\right)\left(t_2 - \bar{Y}\right) \right]^2}{\left[E\left(t_1 - \bar{Y}\right)^2 + E\left(t_2 - \bar{Y}\right)^2 - 2E\left(t_1 - \bar{Y}\right)\left(t_2 - \bar{Y}\right) \right]}. \tag{2.4.18}$$

Using results from (2.4.15) through (2.4.17) in (2.4.18) and simplifying, the mean square error of $t_{33(1)}$ is

$$MSE\left(t_{33(1)}\right) = \theta \bar{Y}^2 C_y^2 \left[1 - \frac{\rho_{xy}^2 + \rho_{yz}^2 - 2\rho_{xy} \rho_{yz} \rho_{xz}}{1 - \rho_{xz}^2} \right],$$

or $$MSE\left(t_{33(1)}\right) = \theta \bar{Y}^2 C_y^2 \left(1 - \rho_{y.xz}^2\right). \tag{2.4.19}$$

From (2.4.15) we can readily conclude that the modified Mohanty (1967) estimator by Samiuddin and Hanif (2006) has a smaller mean square error as compared with the classical regression estimator.

2.4.4 Modification of Mohanty's (1967) Estimator–II

Mohanty (1967) used the regression–in–ratio method to propose his estimator which is given in (2.3.1). The regression part uses regression weight of Y on X in estimation process. Samiuddin and Hanif (2006) argued that the regression weight of Y on X may not be the optimum weight for estimation of population mean. Using this argument Samiuddin and Hanif (2006) proposed the following modified estimator on the lines of Mohanty (1967):

$$t_{34(1)} = \left\{\bar{y} + k_1\left(\bar{X} - \bar{x}\right)\right\}\frac{\bar{Z}}{\bar{z}}. \tag{2.4.20}$$

Using 1.4 in (2.4.20) and simplifying, we can write:

$$t_{34(1)} - \bar{Y} = \bar{e}_y - \frac{\bar{Y}}{\bar{Z}}\bar{e}_z - k_1\bar{e}_x .$$

The mean square error of $t_{34(1)}$ is given as:

$$MSE\left(t_{34(1)}\right) \approx E\left(\bar{e}_y - \frac{\bar{Y}}{\bar{Z}}\bar{e}_z - k_1\bar{e}_x\right)^2. \tag{2.4.21}$$

The optimum value of k_1 which minimizes (2.4.21) is:

$$k_1 = \left[E\left(\bar{e}_x^2\right)\right]^{-1} E\left(\bar{e}_y\bar{e}_x - \frac{\bar{Y}}{\bar{Z}}\bar{e}_z\bar{e}_x\right) = \frac{\bar{Y}}{\bar{X}C_x}\left(\rho_{yx}C_y - \rho_{xz}C_z\right). \tag{2.4.22}$$

Using (2.4.22) in (2.4.21) and simplifying, we can see that the mean square error of $t_{34(1)}$ is:

$$MSE\left(t_{34(1)}\right) = \theta\bar{Y}^2\left[C_y^2 + C_z^2 - 2\rho_{yz}C_yC_z - \left(\rho_{yx}C_y - \rho_{xz}C_z\right)^2\right], \tag{2.4.23}$$

or $$MSE\left(t_{34(1)}\right) = \theta\bar{Y}^2\left[C_y^2\left(1-\rho_{xy}^2\right) + C_z^2\left(1-\rho_{xz}^2\right) - 2C_yC_z\left(\rho_{yz} - \rho_{xy}\rho_{xz}\right)\right]. \tag{2.4.24}$$

Comparing (2.3.2) with (2.4.24) we can see that:

$$MSE\left(t_{30(1)}\right) - MSE\left(t_{34(1)}\right) = C_z^2\rho_{xz}. \tag{2.4.25}$$

From (2.4.25) we can see that $t_{34(1)}$ will be more precise in comparison with $t_{30(1)}$ if correlation between X and Z is positive, otherwise $t_{34(1)}$ will be less precise. Two estimators coincide in precision if $\rho_{xz} = 0$ and in that case optimum value of k given in (2.4.22) will be equal to regression weight of Y on X and in that case two estimators will be exactly equal.

2.4.5 Modification of Mohanty's (1967) Estimator–III

Samiuddin and Hanif (2006) proposed another modification of Mohanty's (1967) estimator which is parallel to (2.4.20) by interchanging the role of the two auxiliary variables. The modified estimator is given as:

$$t_{35(1)} = \left\{\bar{y} + k_2\left(\bar{Z} - \bar{z}\right)\right\}\frac{\bar{X}}{\bar{x}}. \qquad (2.4.26)$$

The optimum value of k_2 which minimizes the mean square of (2.4.26) can be obtained following the procedure of section 2.4.4. The optimum value turned out to be:

$$k_2 = \left[E\left(\bar{e}_z^2\right)\right]^{-1} E\left(\bar{e}_y\bar{e}_z - \frac{\bar{Y}}{\bar{X}}\bar{e}_z\bar{e}_x\right) = \frac{\bar{Y}}{\bar{Z}C_z}\left(\rho_{yz}C_y - \rho_{xz}C_x\right). \qquad (2.4.27)$$

The mean square error of (2.4.26) can be immediately written from (2.4.24) as:

$$MSE\left(t_{35(1)}\right) = \theta_1\bar{Y}^2\left[C_y^2\left(1-\rho_{yz}^2\right) + C_x^2\left(1-\rho_{xz}^2\right) - 2C_yC_x\left(\rho_{xy} - \rho_{yz}\rho_{xz}\right)\right]. \qquad (2.4.28)$$

Further, comparing (2.3.2) with (2.4.27) we can see that:

$$MSE\left(t_{30(1)}\right) - MSE\left(t_{35(1)}\right) = C_x^2\rho_{xz}. \qquad (2.4.29)$$

From (2.4.29) we can see that precision of $t_{35(1)}$ depends on correlation between X and Z as in case of $t_{34(1)}$.

2.4.6 New Modified Estimator

Samiuddin and Hanif (2006) proposed another modified estimator by using weighted average of $t_{34(1)}$ and $t_{35(1)}$. The suggested estimator is:

$$t_{36(1)} = at_{34(1)} + (1-a)t_{35(1)}. \qquad (2.4.30)$$

Rewriting (2.4.30) as:

$$t_{36(1)} - \bar{Y} = \left(t_{35(1)} - \bar{Y}\right) - a\left\{\left(t_{35(1)} - \bar{Y}\right) - \left(t_{34(1)} - \bar{Y}\right)\right\};$$

Squaring and applying expectation we have:

$$MSE\left(t_{36(1)}\right)=E\left[\left(t_{35(1)}-\bar{Y}\right)-a\left\{\left(t_{35(1)}-\bar{Y}\right)-\left(t_{34(1)}-\bar{Y}\right)\right\}\right]^2. \quad (2.4.31)$$

The optimum value of *a*, which minimizes (2.4.31) is:

$$a=\frac{E\left(t_{35(1)}-\bar{Y}\right)^2-E\left(t_{34(1)}-\bar{Y}\right)\left(t_{35(1)}-\bar{Y}\right)}{E\left\{\left(t_{35(1)}-\bar{Y}\right)-\left(t_{34(1)}-\bar{Y}\right)\right\}^2}. \quad (2.4.32)$$

Using (2.4.32) in (2.4.31) and simplifying we have:

$$MSE\left(t_{36(1)}\right)=E\left(t_{35(1)}-\bar{Y}\right)^2-\frac{\left[E\left(t_{35(1)}-\bar{Y}\right)^2-E\left(t_{34(1)}-\bar{Y}\right)\left(t_{35(1)}-\bar{Y}\right)\right]^2}{E\left[\left(t_{34(1)}-\bar{Y}\right)-\left(t_{35(1)}-\bar{Y}\right)\right]^2}$$

or

$$MSE\left(t_{36(1)}\right)=MSE\left(t_{35(1)}\right)-\frac{\left[E\left(t_{35(1)}-\bar{Y}\right)^2-E\left(t_{34(1)}-\bar{Y}\right)\left(t_{35(1)}-\bar{Y}\right)\right]^2}{MSE\left(t_{34(1)}\right)+MSE\left(t_{35(1)}\right)-2E\left(t_{34(1)}-\bar{Y}\right)\left(t_{35(1)}-\bar{Y}\right)} \quad (2.4.33)$$

$MSE\left(t_{34(1)}\right)$ and $MSE\left(t_{35(1)}\right)$ are given in (2.4.24) and (2.4.28) respectively. The quantity $E\left(t_{34(1)}-\bar{Y}\right)\left(t_{35(1)}-\bar{Y}\right)$ is obtained as:

$$\begin{aligned}E\left(t_{34(1)}-\bar{Y}\right)\left(t_{35(1)}-\bar{Y}\right)=\theta\bar{Y}^2\Big[&C_y^2\left(1-\rho_{xy}^2-\rho_{yz}^2+\rho_{xy}\rho_{xz}\rho_{yz}\right)\\&+C_yC_z\left(\rho_{xy}-\rho_{xz}\rho_{yz}\right)\rho_{xz}+C_xC_y\left(\rho_{yz}-\rho_{xy}\rho_{xz}\right)\rho_{xz}-C_xC_z\left(1-\rho_{xz}^2\right)\rho_{xz}\Big]\end{aligned} \quad (2.4.34)$$

The mean square error of $t_{36(1)}$ may be obtained by using (2.4.24), (2.4.28) and (2.4.34) in (2.4.33). The mean square error is not in a compact form.

2.4.7 Modification of Singh and Espejo (2003) Estimator

Hanif et al. (2009) proposed an estimator which was the modification of the Singh and Espejo (2003) estimator:

$$t_{37(1)}=\left\{\bar{y}+k_1\left(\bar{X}-\bar{x}\right)\right\}\left\{k_2\frac{\bar{Z}}{\bar{z}}+\left(1-k_2\right)\frac{\bar{z}}{\bar{Z}}\right\}, \tag{2.4.35}$$

where k_1 and k_2 are constants to be determined so that mean square error of (2.4.35) is minimized. Using notations from 1.4 in (2.4.35) we have:

$$t_{37(1)}-\bar{Y}=\bar{e}_Y-k_1\bar{e}_X+\left(1-2k_2\right)\left(\bar{Y}/\bar{Z}\right)\bar{e}_Z; \tag{2.4.36}$$

Squaring (2.4.36) and applying expectations, the mean square error of $t_{37(1)}$ is given as:

$$\begin{aligned}MSE\left(t_{37(1)}\right)=\Big[&\theta\bar{Y}^2C_Y^2+\theta k_1^2\bar{X}^2C_X^2+\theta\bar{Y}^2C_Z^2+4\theta k_2^2\bar{Y}^2C_Z^2-4\theta k_2\bar{Y}^2C_Z^2\\&-2\theta k_1\bar{X}\bar{Y}C_XC_Y\rho_{XY}-2\theta\bar{Y}^2C_YC_Z\rho_{YZ}-4\theta k_2\bar{Y}^2C_YC_Z\rho_{YZ}\\&-2\theta k_1\bar{X}\bar{Y}C_XC_Z\rho_{XZ}+4\theta k_1k_2\bar{X}\bar{Y}C_XC_Z\rho_{XZ}\Big].\end{aligned} \tag{2.4.37}$$

The optimum values of k_1 and k_2 which minimizes (2.4.37) are:

$$k_1=\frac{S_Y\left(\rho_{XY}-\rho_{XZ}\rho_{YZ}\right)}{S_X\left(1-\rho_{XZ}^2\right)}=\beta_{XY.Z},$$

and

$$k_2=\frac{1}{2}\left\{1+\frac{\bar{Z}}{\bar{Y}}\cdot\frac{S_Y\left(\rho_{YZ}-\rho_{XY}\rho_{XY}\right)}{S_X\left(1-\rho_{XZ}^2\right)}\right\}=\frac{1}{2}\left\{1+\frac{\bar{Z}}{\bar{Y}}\beta_{YZ.X}\right\}.$$

Using optimum values of k_1 and k_2 in (2.4.37) and simplifying, the mean square error of $t_{37(1)}$ is:

$$MSE\left(t_{37(1)}\right)=\theta\bar{Y}^2C_y^2\left(1-\rho_{y.xz}^2\right). \tag{2.4.38}$$

The mean square error given in (2.4.38) is exactly the same as given in (2.4.15).

CHAPTER THREE

GENERALIZED ESTIMATORS FOR SINGLE-PHASE SAMPLING

3.1 Introduction

In Chapter 2 we have discussed several estimators of population mean in single-phase sampling. Many of those estimators are equal in precision to the classical regression estimator to first order. Various generalizations of popular estimators have been proposed in the literature by different survey statisticians. These generalizations have been proposed with a view to increasing precision of estimates. In this Chapter we have focused on generalized estimators for single-phase sampling. In Section 3.2 generalized estimators which use single auxiliary variable will be discussed, while estimators based upon two auxiliary variables will be discussed in Section 3.3.

3.2 Generalized Estimators with One Auxiliary Variable

Suppose a random sample of size n is drawn from a population of N units and information on variable of interest Y is recorded. Suppose further that information on a supplementary variable X is also available. Survey statisticians have proposed a number of generalized estimation strategies on the basis of available information of one auxiliary variable. We discuss some popular generalized estimators as:

3.2.1 Srivastava's (1967) Estimator

Srivastava (1967) proposed the following generalization of the classical ratio estimator:

$$t_{38(1)} = \bar{y}\left(\frac{\bar{X}}{\bar{x}}\right)^{\alpha}, \tag{3.2.1}$$

where α is a constant to be determined. The estimator (3.2.1) reduces to the classical ratio estimator for $\alpha = 1$.

The mean square error of $t_{38(1)}$ may be derived by using $\bar{y} = \bar{Y} + \bar{e}_y$ etc. in (3.2.1) i.e.:

$$t_{38(1)} = \left(\bar{Y} + \bar{e}_y\right)\left(\frac{\bar{X}}{\bar{X} + \bar{e}_x}\right)^{\alpha} = \left(\bar{Y} + \bar{e}_y\right)\left(1 + \frac{\bar{e}_x}{\bar{X}}\right)^{-\alpha}.$$

Expanding and retaining the first term only, we have:

$$t_{38(1)} - \bar{Y} = \bar{e}_y - \alpha\frac{\bar{Y}}{\bar{X}}\bar{e}_x.$$

Squaring and applying expectation, we have:

$$MSE\left(t_{38(1)}\right) \approx \theta_1 \bar{Y}^2\left[C_y^2 + \alpha^2 C_x^2 - 2\alpha C_x C_y \rho_{xy}\right]. \tag{3.2.2}$$

The optimum value of α which minimizes (3.2.2) is $\alpha = \left(\rho C_y\right)/C_x$. Using the value of α in (3.2.2) and simplifying, the mean square error of $t_{38(1)}$ is:

$$MSE\left(t_{38(1)}\right) \approx \theta_1 \bar{Y}^2 C_y^2\left(1 - \rho_{xy}^2\right). \tag{3.2.3}$$

From (3.2.3), we can see that the generalized estimator proposed by Srivastava (1967) is in equivalence with the classical regression estimator as the mean square error of both estimators is same.

3.2.2 Lui's (1990) General Family of Estimators

Lui (1990) proposed a generalized estimator which extends estimators proposed by Searle (1965) and Prasad (1989). The estimator also extends the classical product estimator. The estimator is:

$$t_{39(1)} = \frac{\bar{y}}{\bar{X}}\left(k_1\bar{x} + k_2\right). \tag{3.2.4}$$

The estimator (3.2.4) reduces to Prasad's (1989) estimator for $k_2 = 0$. It reduces to the classical product estimator for $k_1 = 1$ and $k_2 = 0$. We first find optimum values of k_1 and k_2 which minimizes mean square error of $t_{39(1)}$ by writing (3.2.4) as:

$$t_{39(1)}=\frac{\left(\bar{Y}+\bar{e}_y\right)}{\bar{X}}\left\{k_1\left(\bar{X}+\bar{e}_x\right)+k_2\right\}$$

or $$\left(t_{39(1)}-\bar{Y}\right)=\bar{Y}\left[k_1\left(1+\frac{\bar{e}_y}{\bar{Y}}+\frac{\bar{e}_x}{\bar{X}}+\frac{\bar{e}_y\bar{e}_x}{\bar{X}\bar{Y}}\right)+k_2\frac{1}{\bar{X}}\left(1+\frac{\bar{e}_y}{\bar{Y}}\right)-1\right].$$

The mean square error of $t_{39(1)}$ is:

$$MSE\left(t_{39(1)}\right)\approx\bar{Y}^2\left[1+k_1^2A+k_2^2B+2k_1k_2C-2k_1D-2k_2\left(1/\bar{X}\right)\right], \quad (3.2.5)$$

where $A=\left[1+\theta\left(C_y^2+C_x^2+4\rho_{xy}C_xC_y\right)\right]$, $B=\bar{X}^{-2}\left(1+\theta C_y^2\right)$,

$C=\bar{X}^{-1}\left[1+\theta\left(C_y^2+2\rho_{xy}C_xC_y\right)\right]$ and $D=\left(1+\theta\rho_{xy}C_xC_y\right)$.

Differentiating (3.2.5) *w.r.t.* k_1 and k_2 and equating to zero we have:

$$\begin{bmatrix}A & C\\ C & B\end{bmatrix}\begin{bmatrix}k_1\\ k_2\end{bmatrix}=\begin{bmatrix}D\\ \bar{X}^{-1}\end{bmatrix} \quad (3.2.6)$$

Solving (3.2.6) we have:

$$\left.\begin{aligned}k_1&=\frac{\Delta_1}{\Delta}=k_{10}\\ k_2&=\frac{\Delta_2}{\Delta}=k_{20}\end{aligned}\right\}, \quad (3.2.7)$$

where $\Delta=\left(AB-C^2\right)$, (3.2.8)

$$\Delta_1=BD-C\bar{X}^{-1}=\bar{X}^{-1}\left(BD\bar{X}-C\right), \quad (3.2.9)$$

and

$$\Delta_2=A\bar{X}^{-1}-CD=\bar{X}^{-1}\left(A-CD\bar{X}\right). \quad (3.2.10)$$

Using (3.2.7) in (3.2.5) and simplifying, the mean square error of $t_{39(1)}$ is:

$$MSE\left(t_{39(1)}\right)\approx\bar{Y}^2\left[1-\frac{\left(A+BD^2\bar{X}^2-2CD\bar{X}\right)}{\bar{X}^2\left(AB-C^2\right)}\right]. \quad (3.2.11)$$

Das (1988) proposed another estimator which is in equivalence class with the shrinkage regression estimator. The estimator is:

$$t_{40(1)} = W\left[\bar{y} - p\left(\bar{x} - \bar{X}\right)\right]. \tag{3.2.12}$$

The mean square error of (3.2.12) may be readily obtained by using expression (1.6.5) given by Shahbaz and Hanif (2009) which is the same as the mean square error of Lui's (1990) estimator. This shows that Das (1988) and Lui (1990) estimators are always more precise as compared with the classical regression estimator.

3.2.3 Vos's (1980) Generalized Estimator

Vos (1980) proposed the following generalized estimator as the weighted average of mean per unit and Srivastava's (1967) estimator:

$$t_{41(1)} = a\bar{y} + (1-a)\bar{y}\left(\frac{\bar{X}}{\bar{x}}\right)^{p}. \tag{3.2.13}$$

The estimator (3.2.13) reduces to mean per unit estimator for $a = 1$. It reduces to Srivastava's (1967) estimator for $a = 0$. The estimator reduces to Advaryu and Gupta's (1983) estimator for $p = 1$ and it produces the classical ratio estimator as a special case for $a = 0$ and $p = 1$. Using $\bar{y} = \bar{Y} + \bar{e}_y$ etc. we have:

$$t_{41(1)} = a\left(\bar{Y} + \bar{e}_y\right) + (1-a)\left(\bar{Y} + \bar{e}_y\right)\left(1 + \frac{\bar{e}_x}{\bar{X}}\right)^{-p},$$

or $$t_{41(1)} - \bar{Y} = \bar{e}_y - p(1-a)\frac{\bar{Y}}{\bar{X}}\bar{e}_x,$$

or $$t_{41(1)} - \bar{Y} = \bar{e}_y - a_1\frac{\bar{Y}}{\bar{X}}\bar{e}_x \text{ where } a_1 = p(1-a). \tag{3.2.14}$$

Squaring (3.2.14), applying expectation and simplifying, the mean square error of $t_{40(1)}$ turned out to be:

$$MSE\left(t_{41(1)}\right) \approx \theta_1 \bar{Y}^2 C_y^2\left(1 - \rho_{xy}^2\right). \tag{3.2.15}$$

The mean square error of the Vos (1980) estimator, given in (3.2.15), shows that it is also in equivalence class with the classical regression estimator.

3.2.4 Naik and Gupta (1991) General Class of Estimators

Naik and Gupta (1991) proposed a general class of estimators for estimating population mean using information of a single auxiliary variable. The estimator proposed by Naik and Gupta (1991) is:

$$t_{42(1)} = \bar{y}\left[\left\{\bar{X} + a\left(\bar{x} - \bar{X}\right)\right\} / \left\{\bar{X} + b\left(\bar{x} - \bar{X}\right)\right\}\right]^{p}, \tag{3.2.16}$$

where p, a and b are constants. The estimator (3.2.16) produces several estimators as special cases for various choices of constants a, b and p. A list of special cases of $t_{42(1)}$ is given in table 3.1. Using (3.2.16) we have:

$$t_{42(1)} = \left(\bar{Y} + \bar{e}_y\right)\left(\frac{\bar{X} + a\bar{e}_x}{\bar{X} + b\bar{e}_x}\right)^{p} = \left(\bar{Y} + \bar{e}_y\right)\left(1 + \frac{a}{\bar{X}}\bar{e}_x\right)^{p}\left(1 + \frac{b}{\bar{X}}\bar{e}_x\right)^{-p}.$$

Expanding and retaining linear terms only we have:

$$t_{42(1)} = \left(\bar{Y} + \bar{e}_y\right)\left(1 + \frac{ap}{\bar{X}}\bar{e}_x\right)\left(1 - \frac{bp}{\bar{X}}\bar{e}_x\right),$$

or $t_{42(1)} - \bar{Y} = \bar{e}_y - p\left(b - a\right)\dfrac{\bar{Y}}{\bar{X}}\bar{e}_x$,

or $t_{42(1)} - \bar{Y} = \bar{e}_y - p_1\bar{e}_x$ with $p_1 = p\left(b - a\right)\dfrac{\bar{Y}}{\bar{X}}$. (3.2.17)

Squaring (3.2.17), applying expectation and simplifying we have:

$$MSE\left(t_{42(1)}\right) \approx \theta_1\left(\bar{Y}^2C_y^2 + p_1^2\bar{X}^2C_x^2 - 2p_1\bar{X}\bar{Y}\rho_{xy}C_xC_y\right). \tag{3.2.18}$$

The optimum value of p_1 which minimizes (3.2.18) is $p_1 = \left(\rho_{xy}\bar{Y}C_y\right) / \left(\bar{X}C_x\right)$. Using this value of p_1 in (3.2.18) and simplifying, the mean square error of $t_{42(1)}$ is:

$$MSE\left(t_{42(1)}\right) \approx \theta_1\bar{Y}^2C_y^2\left(1 - \rho_{xy}^2\right). \tag{3.2.19}$$

The mean square error of $t_{42(1)}$, given in (3.2.19), shows that Naik and Gupta (1991) is in equivalence class with the classical regression estimator.

3.2.5 Das and Tripathi (1980) General Estimator

The estimator proposed by Das and Tripathi (1980) is:

$$t_{43(1)} = \frac{\bar{y} - p\left(\bar{x} - \bar{X}\right)}{\left[\bar{x} - q\left(\bar{x} - \bar{X}\right)\right]^{\alpha}} \bar{X}^{\alpha}. \qquad (3.2.20)$$

The estimator (3.2.20) produces several estimators as a special case. For example the estimator reduces to mean per unit estimator for $p = 0$ and $\alpha = 0$. It reduces to Srivastava's (1967) estimator for $p = q = 0$. The estimator produces a difference estimator as a special case for $\alpha = 0$. Using notations from section 1.4 in (3.2.20) we have:

$$t_{43(1)} = \frac{\bar{Y} + \bar{e}_y - p\bar{e}_x}{\left[\bar{X} + (1-q)\bar{e}_x\right]^{\alpha}} \bar{X}^{\alpha} = \left(\bar{Y} + \bar{e}_y - p\bar{e}_x\right)\left\{1 - \alpha(1-q)\frac{\bar{e}_x}{\bar{X}}\right\},$$

or $t_{43(1)} - \bar{Y} = \bar{e}_y - q_1\bar{e}_x$ with $q_1 = p + \alpha(1-q)\frac{\bar{Y}}{\bar{X}}$. (3.2.21)

Comparing (3.2.21) with (3.2.17), we can readily see that mean square error of $t_{43(1)}$ is the same as the mean square error of the classical regression estimator. We, therefore, say that the Das and Tripathi (1980a) estimator is also in equivalence class with the classical regression estimator.

Table 1: Special cases of $t_{42(1)}$

S#	Value of Constant	Estimator	MSE after putting different value of p, a and b
1	p=0,	$t_0 = \bar{y}$. (mean per unit)	$\theta_1 \bar{Y} C_y^2 = Var(t_0)$.
2	p=1, a=0, b=1	$t_{1(1)} = \frac{\bar{y}}{\bar{x}} \bar{X}$. (Ratio)	$\theta_1 \bar{Y}^2 \left(C_y^2 - 2C_x C_y \rho_{xy} + C_x^2\right)$.
3	p=1, a=1, b=0	$t_{3(1)} = \bar{y} \frac{\bar{x}}{\bar{X}}$. (Product)	$\theta_1 \bar{Y}^2 \left[C_y^2 + 2C_y C_x \rho_{xy} + C_x^2\right]$.
4	p=α, a=0, b=1	$t_{4(1)} = \bar{y}\left(\frac{\bar{X}}{\bar{x}}\right)^{\alpha}$. (Srivastava -1967)	$\theta_1 \bar{Y}^2 C_y^2 \left(1-\rho_{xy}^2\right)$.
5	p=1, a=0, b=b	$t_{5(1)} = \bar{y}\left[\frac{\bar{X}}{\bar{X} + b\left(\bar{x} - \bar{X}\right)}\right]$. (Walsh-1970; Reddy -1973-74)	$\theta_1 \bar{Y}^2 C_y^2 \left(1-\rho_{xy}^2\right)$.
6	p=1, a=(1-w), b=w	$t_{6(1)} = \bar{y}\left[\frac{w\bar{X} + (1-w)\bar{x}}{w\bar{x} + (1-w)\bar{X}}\right]$. (Sahai-1979)	$\theta_1 \bar{Y}^2 C_y^2 \left(1-\rho_{xy}^2\right)$.
7	p=1, a=a, b=1	$t_{7(1)} = a\bar{y} + (1-a)\frac{\bar{y}}{\bar{x}} \bar{X}$. (Vos-1980, Advaryu and Gupta-1983)	$\theta_1 \bar{Y}^2 C_y^2 \left(1-\rho_{xy}^2\right)$.
8	p=1, a=a, b=0	$t_{38(1)} = a\bar{y} + (1-a)\bar{y}\left(\frac{\bar{X}}{\bar{x}}\right)$. (Vos-1980, Adhvaryu and Gupta-1983)	$\theta_1 \bar{Y}^2 C_y^2 \left(1-\rho_{xy}^2\right)$.

3.2.6 Tripathi, Das and Khare (1994) General Class of Estimators

Tripathi et al. (1994) provided a further generalization of the Das and Tripathi (1980) estimator. The estimator proposed by Tripathi et al. (1994) is given as:

$$t_{44(1)} = \frac{\bar{y} - p\left(\bar{x}^{\beta} - \bar{X}^{\beta}\right)}{\left[\bar{x} - q\left(\bar{x}^{\beta} - \bar{X}^{\beta}\right)\right]^{\alpha}} \bar{X}^{\alpha}. \tag{3.2.22}$$

The estimator (3.2.22) reduces to Das and Tripathi (1980) estimator for $\beta = 1$. Other special cases of estimator (3.2.22) are given in table 3.2. Using $\bar{y} = \bar{Y} + \bar{e}_y$ etc., the estimator (3.2.22) can be written as:

$$t_{44(1)} = \frac{\left(\bar{Y} + \bar{e}_y\right) - p\left[\left(\bar{X} + \bar{e}_x\right)^{\beta} - \bar{X}^{\beta}\right]}{\left[\bar{X} + \bar{e}_x - \left\{q\left(\bar{X} + \bar{e}_x\right)^{\beta} - \bar{X}^{\beta}\right\}\right]^{\alpha}} \bar{X}^{\alpha}.$$

Expanding up to $O(1/n)$ and on simplification we get:

$$t_{44(1)} = \left[\left(\bar{Y} + \bar{e}_y\right) - p\left(\beta \bar{X}^{\beta-1} \bar{e}_x\right)\right]\left[1 + \left(\bar{e}_x - q\beta \bar{X}^{\beta-1} \bar{e}_x\right)/\bar{X}\right]^{-\alpha},$$

or $$t_{44(1)} = \left[\bar{Y} + \bar{e}_y - p\beta \bar{X}^{\beta-1} \bar{e}_x\right]\left[1 - \alpha \frac{\bar{e}_x}{\bar{X}}\left(1 - q\beta \bar{X}^{\beta-1}\right)\right],$$

or $$t_{44(1)} - \bar{Y} = \bar{e}_y - \left\{\alpha \frac{\bar{Y}}{\bar{X}}\left(1 - q\beta \bar{X}^{\beta-1}\right) + p\beta \bar{X}^{\beta-1}\right\} \bar{e}_x,$$

or $$t_{44(1)} - \bar{Y} = \bar{e}_y - A\bar{e}_x \text{ where } A = \left\{\alpha \frac{\bar{Y}}{\bar{X}}\left(1 - q\beta \bar{X}^{\beta-1}\right) + p\beta \bar{X}^{\beta-1}\right\}. \tag{3.2.23}$$

Squaring (3.2.23), applying expectation and simplifying, the mean square error of $t_{44(1)}$ may be written as:

$$MSE\left(t_{44(1)}\right) \approx \theta_1\left[\bar{Y}^2 C_y^2 + A^2 C_x^2 \bar{X}^2 - 2A^2 \overline{YX}\, C_x C_y \rho_{xy}\right]. \tag{3.2.24}$$

Comparing (3.2.24) with (2.2.11) we can see that the mean square error of the Tripathi et al. (1994) estimator is the same as the mean square error of the classical regression estimator. For various values of constants special cases are given in Table 3.2.

Table 3.2: Special cases of this general class of estimator $t_{44(1)}$

S#	p	q	α	β	Estimator	Reference
1	0	*0*	α	β	$\bar{y}\left(\frac{\bar{X}}{\bar{x}}\right)^{a}$.	Srivastava (1967)
2	p	q	0	1	$\bar{y}-p\left(\bar{x}-\bar{X}\right)$.	Tripathi (1970,80)
3	p	q	1	1	$\frac{\bar{y}-p\left(\bar{x}-\bar{X}\right)}{\bar{x}-q\left(\bar{x}-\bar{X}\right)}\bar{X}$.	Tripathi (1970,80)
4	0	*1-w*	1	1	$\frac{\bar{y}}{\bar{x}-(1-w)\left(\bar{x}-\bar{X}\right)}\bar{X}$.	Singh(1969), Reddy(1974)
5	$w\left(\frac{\bar{y}}{\bar{X}}\right)$	q	0	1	$\bar{y}-w\frac{\bar{y}}{\bar{X}}\left(\bar{x}-\bar{X}\right)$.	Rao (1968), Kumar (1980), Vos(1980)
6	$(1-w)\frac{\bar{y}}{\bar{X}}$	*0*	1	1	$\left\{\bar{y}-(1-w)\frac{\bar{y}}{\bar{X}}\left(\bar{x}-\bar{X}\right)\right\}\frac{\bar{X}}{\bar{x}}$.	Gupta (1978)
7	$(1-w)\frac{\bar{y}}{\bar{X}}$	*0*	-1	1	$\bar{y}\left[1-(1-w)\frac{1}{\bar{X}}\left(\bar{x}-\bar{X}\right)\right]\frac{\bar{x}}{\bar{X}}$.	Gupta (1978)
8	$-(1-w)\frac{\bar{y}}{\bar{X}}$	q	0	1	$\bar{y}+(1-w)\frac{\bar{y}}{\bar{X}}\left(\bar{x}-\bar{X}\right)$.	Ray, et al. (1979), Vos (1980)

9	$w\lambda$	q	0	1	$\bar{y}-w\lambda\left(\bar{x}-\bar{X}\right).$	Das (1988)
10	$w\beta$	q	0	1	$\bar{y}-w\beta\left(\bar{x}-\bar{X}\right).$	Das (1988)
11	$\frac{1}{2}\frac{\bar{y}}{\bar{x}^{\beta}}$	0	1+w	-2	$\bar{y}\left[1-\frac{1}{2}\left\{1-\left(\frac{\bar{x}}{\bar{X}}\right)^2\right\}\right]\left(\frac{\bar{X}}{\bar{x}}\right)^{1+w}.$	Ray and Sahai (1978)
12	$-(1-w)\frac{\bar{y}}{\bar{X}}$	1−w	1	1	$\frac{\bar{y}\left[1+(1-w)\frac{1}{\bar{X}}\left(\bar{x}-\bar{X}\right)\right]}{\bar{x}-(1-w)\left(\bar{x}-\bar{X}\right)}\bar{X}.$	Sahai (1979)
13	$\frac{1}{1-f}\frac{\bar{y}}{\bar{x}}$	0	-1	1	$\bar{y}\left[1-\frac{1}{(1-f)\bar{x}}\left(\bar{x}-\bar{X}\right)\right]\frac{\bar{x}}{\bar{X}}.$	Srivenkataramana and Srinath (1976) Srivenkataramana (1980)
14	$f.\left(\frac{\bar{y}}{\bar{X}}\right)f=\frac{n}{N}$	q	0	1	$\bar{y}\left[1-\frac{f}{\bar{X}}\left(\bar{x}-\bar{X}\right)\right].$	Srivenkataramana and Tracy (1979)
15	$-f\left(\frac{\bar{y}}{\bar{x}}\right)$	q	0	1	$\bar{y}\left[1+\frac{f}{\bar{X}}\left(\bar{x}-\bar{X}\right)\right].$	Srivenkataramana and Tracy (1979)
16	$-(1-f)\frac{\bar{y}}{\bar{X}}$	q	0	1	$\bar{y}\left[1+\frac{1-f}{\bar{X}}\left(\bar{x}-\bar{X}\right)\right].$	Srivenkataramana and Tracy (1979)
17	$\frac{\bar{y}}{w-\bar{X}}$	q	0	1	$\bar{y}\left[1-\frac{1}{w-\bar{X}}\left(\bar{x}-\bar{X}\right)\right].$	Srivenkataramana and Tracy (1980), Kaur (1983)

18	$\frac{\bar{y}}{\bar{X}^{\beta}}$	q	0	β	$\bar{y}\left[1-\frac{1}{\bar{X}^{\beta}}\left(\bar{x}^{\beta}-\bar{X}^{\beta}\right)\right].$	Sahai and Ray (1980)
19	$-\frac{\bar{y}}{w+\bar{X}}$	q	0	1	$\bar{y}\left[1+\frac{1}{w+\bar{X}}\left(\bar{x}-\bar{X}\right)\right].$	Srivenkataramana and Tracy (1981) Shukla and Pandey (1982)
20	$w\frac{\bar{y}}{\bar{x}^{\beta}}$	q	0	β	$\bar{y}\left[1-\frac{w}{\bar{x}^{\beta}}\left(\bar{x}^{\beta}-\bar{X}^{\beta}\right)\right].$	Sisodia and Dwivedi (1981a)
21	$\frac{-\left(t_2-1\right)\bar{y}}{\bar{x}+\left(t_1+t_2-1\right)\bar{X}}$	q	0	1	$\bar{y}\left[1+\frac{\left(t_1-1\right)}{\bar{x}+\left(t_1+t_2-1\right)\bar{X}}\left(\bar{x}-\bar{X}\right)\right].$	Ray and Sahai (1980)
22	$\frac{\left(t_2-1\right)\bar{y}}{\bar{X}+\left(t_1+t_2-1\right)\bar{x}}$	q	0	1	$\bar{y}\left[1-\frac{\left(t_2-1\right)}{\bar{X}+\left(t_1+t_2-1\right)\bar{x}}\left(\bar{x}-\bar{X}\right)\right].$	Ray and Sahai (1980)
23	-1	0	1	$\frac{1}{2}$	$\left[\bar{y}+\left(\bar{x}^{\frac{1}{2}}-\bar{X}^{\frac{1}{2}}\right)\right]\frac{\bar{X}}{\bar{x}}.$	Ray and Singh (1981)
24	-1	0	1	$-\frac{1}{2}$	$\bar{y}+\left[\bar{x}^{-\frac{1}{2}}-\bar{X}^{-\frac{1}{2}}\right]\frac{\bar{X}}{\bar{x}}.$	Ray and Singh (1981)
25	$w\frac{\bar{y}}{\bar{X}^{\beta}}$	q	0	β	$\bar{y}\left[1-\frac{w}{\bar{X}^{\beta}}\left(\bar{x}^{\beta}-\bar{X}^{\beta}\right)\right].$	Singh and Pandey (1983)

26	$-\frac{(1-f)(t_2-1)\bar{y}}{\bar{x}+(t_1+t_2-1)\bar{X}}$	q	0	1	$\bar{y}\left[1+\frac{(1-f)(t_2-1)}{\bar{x}+(t_1+t_2-1)\bar{X}}(\bar{x}-\bar{X})\right].$	Upadhyaya and Singh (1984)
27	$\frac{-(t_2-1)\bar{y}}{\bar{x}^*+(t_1+t_2-1)\bar{X}^*}\cdot\frac{\bar{x}^*-\bar{X}}{\bar{x}-\bar{X}}$ $\bar{x}^*=\frac{N\bar{X}-n\bar{x}}{N-n}$	q	0	1	$\bar{y}\left[1+\frac{\left\{(t_2-1)\left(\frac{N\bar{X}-n\bar{x}}{N-n}-\bar{X}\right)\right\}}{\left\{\frac{N\bar{X}-n\bar{x}}{N-n}+(t_2+t_1-1)\bar{X}\right\}}\right].$	Upadhyaya and Singh (1984)
28	$\frac{(t_2-1)\bar{y}}{\bar{X}+(t_1+t_2-1)\bar{x}^*}\ \frac{\bar{x}^*-\bar{X}}{\bar{x}-\bar{X}}$	q	0	1	$\bar{y}\left[1-\frac{\left\{(t_2-1)\left(\frac{N\bar{X}-n\bar{x}}{N-n}-\bar{X}\right)\right\}}{\left\{\bar{X}+(t_2+t_1-1)\frac{N\bar{X}-n\bar{x}}{N-n}\right\}}\right].$	Upadhyaya and Singh (1984)
29	$-\frac{(1-f)\bar{y}}{\bar{X}-f\bar{x}}$	q	0	1	$\bar{y}\left[1+\frac{1-f}{\bar{X}-f\bar{x}}(\bar{x}-\bar{X})\right].$	Srivastava (1983)
30	$\frac{\bar{y}}{w+\bar{x}}$	q	0	1	$\bar{y}\left[1-\frac{1}{w+\bar{x}}(\bar{x}-\bar{X})\right].$	Kulkarni (1978)
31	$-w\left(\frac{\bar{y}}{\bar{X}}\right)$	q	0	1	$\bar{y}\left[1+\frac{w}{\bar{X}}(\bar{x}-\bar{X})\right].$	Chaubey, Singh and Dwivedi (1984)

3.3 Generalized Estimators with Two Auxiliary Variables

We have discussed certain generalized estimators for single-phase sampling which are based upon single auxiliary variable. Various survey statisticians have proposed a number of estimators in single-phase sampling which are based upon information of two auxiliary variables. We have given a detailed discussion of single-phase sampling estimators with two auxiliary variables in section 2.3. Generalizations of estimators discussed in that section can be readily seen as generalizations discussed in section 3.2. Samiuddin and Hanif (2006) have been very instrumental in development of generalized estimators for single-phase sampling using information of two auxiliary variables. In this section we present some estimators for single-phase sampling which are based upon information of two auxiliary variables.

3.3.1 Generalized Chain Ratio Estimator–I

The chain type ratio estimator has been an area of study by various survey statisticians. Swain (1970) and Chand (1975) use a chain ratio estimator in two phase sampling. Samiuddin and Hanif (2006) used the same idea in single-phase sampling when population information of two auxiliary variables X and Z is available in addition to sample information of these auxiliary variables and variable of interest Y.

Samiuddin and Hanif (2006) have argued that information of one additional auxiliary variable can be used to increase precision of estimates. Using this argument they proposed the following generalized chain ratio estimator:

$$t_{45(1)} = \bar{y}\left(\frac{\bar{X}}{\bar{x}}\right)^{\alpha}\left(\frac{\bar{Z}}{\bar{z}}\right)^{1-\alpha}. \tag{3.3.1}$$

The estimator (3.3.1) is equivalent to the classical ratio estimator with auxiliary variable X for $\alpha = 1$ and it is equal to the classical ratio estimator with auxiliary variable Z for $\alpha = 0$. Using $\bar{y} = \bar{Y} + \bar{e}_y$ etc. in (3.3.1) we have:

$$t_{45(1)} = \left(\bar{Y} + \bar{e}_y\right)\left(\frac{\bar{X}}{\bar{X} + \bar{e}_x}\right)^{\alpha}\left(\frac{\bar{Z}}{\bar{Z} + \bar{e}_z}\right)^{1-\alpha},$$

or $$t_{45(1)} = \left(\bar{Y} + \bar{e}_y\right)\left(1 + \frac{\bar{e}_x}{\bar{X}}\right)^{-\alpha}\left(1 + \frac{\bar{e}_z}{\bar{Z}}\right)^{-(1-\alpha)}.$$

Expanding the power series and retaining linear terms only we can write above equation as:

$$t_{45(1)}-\overline{Y}=\left(\overline{e}_y-\frac{\overline{Y}}{\overline{\overline{Z}}}\overline{e}_z\right)-\alpha\overline{Y}\left(\frac{\overline{e}_x}{\overline{\overline{X}}}-\frac{\overline{e}_z}{\overline{\overline{Z}}}\right),$$

Squaring and applying expectation the mean square error of $t_{45(1)}$ is:

$$MSE\left(t_{45(1)}\right)=E\left[\left(\overline{e}_y-\frac{\overline{Y}}{\overline{\overline{Z}}}\overline{e}_z\right)-\alpha\overline{Y}\left(\frac{\overline{e}_x}{\overline{\overline{X}}}-\frac{\overline{e}_z}{\overline{\overline{Z}}}\right)\right]^2. \qquad (3.3.2)$$

The optimum value of α which minimizes (3.3.2) is:

$$\alpha=E\left[\left(\overline{e}_y-\frac{\overline{Y}}{\overline{\overline{Z}}}\overline{e}_z\right)\left(\frac{\overline{e}_x}{\overline{\overline{X}}}-\frac{\overline{e}_z}{\overline{\overline{Z}}}\right)\right]\left[\overline{Y}E\left(\frac{\overline{e}_x}{\overline{\overline{X}}}-\frac{\overline{e}_z}{\overline{\overline{Z}}}\right)^2\right]^{-1}, \qquad (3.4.3)$$

Expanding (3.3.2) and using (3.3.3), the mean square error of $t_{45(1)}$ can be written as:

$$MSE\left(t_{45(1)}\right)=E\left(\overline{e}_y-\frac{\overline{Y}}{\overline{\overline{Z}}}\overline{e}_z\right)^2-\frac{\left[E\left(\overline{e}_y-\frac{\overline{Y}}{\overline{\overline{Z}}}\overline{e}_z\right)\left(\frac{\overline{e}_x}{\overline{\overline{X}}}-\frac{\overline{e}_z}{\overline{\overline{Z}}}\right)\right]^2}{E\left(\frac{\overline{e}_x}{\overline{\overline{X}}}-\frac{\overline{e}_z}{\overline{\overline{Z}}}\right)^2}.$$

Expanding squares and cross products, applying expectations and simplifying we obtain the following expression for mean square error of $t_{45(1)}$:

$$MSE\left(t_{45(1)}\right)\approx\theta_1\overline{Y}^2\left[\left(C_y^2+C_z^2-2\rho_{yz}C_yC_z\right)-\frac{\left(\rho_{yx}C_yC_x-C_yC_z\rho_{yz}-C_zC_x\rho_{xz}+C_z^2\right)^2}{C_x^2+C_z^2-2\rho_{xz}C_xC_z}\right]. \qquad (3.3.4)$$

The mean square error given in (3.3.4) is the same as the mean square error of the revised ratio estimator given in (2.4.6). We can therefore comment that the revised ratio estimator and generalized chain ratio estimator, both proposed by Samiuddin and Hanif (2006), are in equivalence class.

3.3.2 Generalized Chain Ratio Type Estimator-II

Samiuddin and Hanif (2006) have proposed another generalized chain ratio type estimator. The proposed estimator is:

$$t_{46(1)} = \bar{y}\left(\frac{\bar{X}}{\bar{x}}\right)^{\alpha}\left(\frac{\bar{Z}}{\bar{z}}\right)^{\beta}. \quad (3.3.5)$$

The estimator (3.3.5) is equal to generalized chain ratio estimator–I of Samiuddin and Hanif (2006) for $\beta = 1-\alpha$. The estimator (3.3.5) is different from (3.3.1) in that the sum of powers need not be 1. The estimator (3.3.5) produces the Srivastava (1967) estimator based on auxiliary variable *X* for $\beta = 0$. It produces the Srivastava (1967) estimator based on auxiliary variable *Z* for $\alpha = 0$. It reduces to Classical ratio estimator for *X* for $\alpha = 1$ and $\beta = 0$. Similarly the classical ratio estimator for *Z* for $\alpha = 0$ and $\beta = 1$. The mean square error of (3.3.5) may be derived as below:

Using $\bar{y} = \bar{Y} + \bar{e}_y$ etc. in (3.3.5) we have:

$$t_{46(1)} = \left(\bar{Y} + \bar{e}_y\right)\left(1 + \frac{\bar{e}_x}{\bar{X}}\right)^{-\alpha}\left(1 + \frac{\bar{e}_z}{\bar{Z}}\right)^{-\beta}.$$

Expanding the above equation and ignoring square and higher terms we have:

$$t_{46(1)} - \bar{Y} = \bar{e}_y - \alpha\frac{\bar{Y}}{\bar{X}}\bar{e}_x - \beta\frac{\bar{Y}}{\bar{Z}}\bar{e}_z.$$

The mean square error of $t_{46(1)}$ is:

$$MSE\left(t_{46(1)}\right) = E\left(\bar{e}_y - \alpha\frac{\bar{Y}}{\bar{X}}\bar{e}_x - \beta\frac{\bar{Y}}{\bar{Z}}\bar{e}_z\right)^2. \quad (3.3.6)$$

We first obtain optimum values of α and β which minimizes (3.3.6). Partially differentiating (3.3.6) *w.r.t.* α and setting the derivatives to zero we have:

$$E\left(\bar{e}_y\bar{e}_x - \alpha\frac{\bar{Y}}{\bar{X}}\bar{e}_x^2 - \beta\frac{\bar{Y}}{\bar{Z}}\bar{e}_z\bar{e}_x\right) = 0,$$

or $\alpha C_x + \beta\rho_{xz}C_z = \rho_{yx}C_y$. (3.3.7)

Again differentiating (3.4.6) *w.r.t.* β and setting the derivative to zero we have:

$$E\left(\bar{e}_y\bar{e}_z-\alpha\frac{\bar{Y}}{\bar{X}}\bar{e}_x\bar{e}_z-\beta\frac{\bar{Y}}{\bar{Z}}\bar{e}_z^2\right)=0,$$

or $\alpha\rho_{xz}C_x+\beta C_z=\rho_{yz}C_y$. (3.3.8)

Solving (3.3.7) and (3.3.8) simultaneously, we have the following optimum values of α and β:

$$\alpha=\frac{C_y}{C_x}\cdot\frac{\rho_{yx}-\rho_{yz}\rho_{xz}}{1-\rho_{xz}^2}=\frac{\bar{Y}}{\bar{X}}\beta_{yx.z}, \quad (3.3.9)$$

and

$$\beta=\frac{C_y}{C_z}\cdot\frac{\rho_{yz}-\rho_{yx}\rho_{xz}}{1-\rho_{xz}^2}=\frac{\bar{Y}}{\bar{Z}}\beta_{yz.x}, \quad (3.3.10)$$

where $\beta_{yx.z}$ and $\beta_{yz.x}$ are partial regression coefficients. Using (3.3.9) and (3.3.10) in (3.3.6) and simplifying, the mean square error of $t_{46(1)}$ is:

$$MSE\left(t_{46(1)}\right)\approx\theta_1\bar{Y}^2C_y^2\left[\frac{1-\rho_{xz}^2-\rho_{yx}^2-\rho_{yz}^2+2\rho_{xy}\rho_{xz}\rho_{yz}}{1-\rho_{xz}^2}\right],$$

or $MSE\left(t_{46(1)}\right)\approx\theta_1\bar{Y}^2C_y^2\left(1-\rho_{y.xz}^2\right)$. (3.3.11)

Comparing (3.3.11) with (2.4.19) and with (2.4.38) we can see that the generalized chain ratio estimator–II, and modified Singh and Espejo (2003) estimator of Samiuddin and Hanif (2006) are in equivalence class.

3.3.3 Modification of Mohanty (1967) Estimator

We have discussed a modification of the Mohanty (1967) estimator given by Samiuddin and Hanif (2006) in section 2.4.4. Samiuddin and Hanif (2006) have also proposed a general modification of the Mohanty (1967) estimator which is given as:

$$t_{47(1)}=\left\{\bar{y}+k\left(\bar{X}-\bar{x}\right)\right\}\left(\frac{\bar{Z}}{\bar{z}}\right)^{\alpha}. \quad (3.3.12)$$

Replacing $\bar{y}$ with $\bar{Y}+\bar{e}_y$ etc., we can write (3.3.12) as:

$$t_{47(1)} = \left\{\left(\bar{Y}+\bar{e}_y\right)-k\bar{e}_x\right\}\left(1+\frac{\bar{e}_z}{\bar{Z}}\right)^{-\alpha}.$$

Expanding and ignoring higher order terms we have:

$$t_{47(1)} - \bar{Y} = \bar{e}_y - k\bar{e}_x - \alpha\frac{\bar{Y}}{\bar{Z}}\bar{e}_z.$$

The mean square error is therefore:

$$MSE\left(t_{47(1)}\right) = E\left(\bar{e}_y - k\bar{e}_x - \alpha\frac{\bar{Y}}{\bar{Z}}\bar{e}_z\right)^2. \tag{3.3.13}$$

Optimum values of k and α which minimizes (3.3.13) can be immediately obtained as $k = \beta_{yx.z}$ and $\alpha = \beta_{yz.x}$. Substituting the optimum values of k and α in (3.3.13) and simplifying, the mean square error of $t_{47(1)}$ is exactly same as the mean square error of $t_{46(1)}$.

3.4 Some New Aspects of Use of Auxiliary Information

We have discussed estimators which use information on two auxiliary variables in previous pages. In discussion of these estimators we have seen that all those estimators use information about population characteristics of auxiliary variables. Often population information of auxiliary variables is unavailable and we still want to use that information in the estimation process. Samiuddin and Hanif (2006) have discussed various possibilities of availability of auxiliary information when two additional variables are used in the estimation process. These possibilities are:

a) Population Information of both auxiliary variables is available; Samiuddin and Hanif (2006) called it **Full Information Case** (FIC)
b) Population Information of only one auxiliary variable is available; referred as **Partial Information Case** (PIC)
c) Population Information of none of the auxiliary variables is available; referred as **No Information Case** (NIC)

In the following we present regression and ratio estimators discussed by Samiuddin and Hanif (2006) under the above situations.

3.4.1 Regression Estimators

We have seen in previous Chapter that a regression estimator is simpler and more efficient as compared with a ratio estimator. Extension of regression estimator when information on several auxiliary variables is available is relatively simple. Samiuddin and Hanif (2006) proposed three versions of regression estimator based upon information of two auxiliary variables. We discuss these versions in the following:

3.4.1.1 Regression Estimator for No Information Case

Samiuddin and Hanif (2006) proposed a regression estimator when population information of auxiliary variables is not available. The estimator is:

$$t_{48(1)} = \overline{y} + \alpha\overline{x} + \beta\overline{z}. \quad (3.4.1)$$

The estimator (3.4.1) is a biased estimator with $Bias\left(t_{48(1)}\right) = \alpha\overline{X} + \beta\overline{Z}$. The mean square error of (3.4.1) can be obtained by using (3.4.5). We can readily show that variance of $t_{48(1)}$ is:

$$MSE\left(t_{48(1)}\right) = \theta_1\overline{Y}^2C_y^2\left(1-\rho_{y.xz}^2\right).$$

The mean square error is, therefore,

$$MSE\left(t_{48(1)}\right) \approx \theta_1\overline{Y}^2C_y^2\left(1-\rho_{y.xz}^2\right) + \beta_{yx.z}^2\overline{X}^2 + \beta_{yz.x}^2\overline{Z}^2. \quad (3.4.2)$$

3.4.1.2 Regression Estimator for Partial Information Case

Samiuddin and Hanif (2006) proposed the following regression estimator when population information of one auxiliary variable is available and for the other auxiliary variable the population information is unavailable:

$$t_{49(1)} = \overline{y} + \alpha\left(\overline{X} - \overline{x}\right) + \beta\overline{z}. \quad (3.4.3)$$

We can readily see that (3.4.3) is a biased estimator with $Bias\left(t_{49(1)}\right) = \beta\overline{Z}$. Samiuddin and Hanif (2006) have argued that the mean square error of (3.4.3.) can be obtained as:

$$MSE\left(t_{49(1)}\right) = Var\left(t_{49(1)}\right) + \left\{Bias\left(t_{49(1)}\right)\right\}^2. \quad (3.4.4)$$

Now

$$Var\left(t_{49(1)}\right)=E\left\{t_{49(1)}-E\left(t_{49(1)}\right)\right\}^2=E\left\{\left(\bar{y}-\bar{Y}\right)+\alpha\left(\bar{X}-\bar{x}\right)-\beta\left(\bar{Z}-\bar{z}\right)\right\}^2$$

Using $\bar{y}=\bar{Y}+\bar{e}_y$ etc., we have:

$$Var\left(t_{49(1)}\right)=E\left(\bar{e}_y-\alpha\bar{e}_x+\beta\bar{e}_z\right)^2. \tag{3.4.5}$$

Optimum values of α and β which minimizes (3.4.5) are $\alpha=\beta_{yx.z}$ and $\beta=-\beta_{yz.x}$. Using optimum values of α and β in (3.4.5) and simplifying, we have:

$$Var\left(t_{49(1)}\right)=\theta_1\bar{Y}^2C_y^2\left(1-\rho_{y.xz}^2\right). \tag{3.4.6}$$

Using (3.4.6) and value of bias in (3.4.4), the mean square error of $t_{49(1)}$ is:

$$MSE\left(t_{49(1)}\right)\approx\theta_1\bar{Y}^2C_y^2\left(1-\rho_{y.xz}^2\right)+\beta_{yz.x}^2\bar{Z}^2, \tag{3.4.7}$$

where $\beta_{yz.x}$ is given above. Comparison of (3.4.8) with (3.4.3) immediately shows that $t_{48(1)}$ is more precise as compared with $t_{49(1)}$. Samiuddin and Hanif (2006) have also commented that when population information for auxiliary variable Z is available and for X is unavailable then analogously the estimator is:

$$t_{50(1)}=\bar{y}+\alpha\bar{x}+\beta\left(\bar{Z}-\bar{z}\right),$$

with $MSE\left(t_{50(1)}\right)=\theta_1\bar{Y}^2C_y^2\left(1-\rho_{y.xz}^2\right)+\beta_{yx.z}^2\bar{X}^2$.

3.4.1.3 Regression Estimator for Full Information Case

The regression estimator for FIC, proposed by Samiuddin and Hanif (2006) is:

$$t_{50(1)}=\bar{y}+\alpha\left(\bar{X}-\bar{x}\right)+\beta\left(\bar{Z}-\bar{z}\right). \tag{3.4.8}$$

Using $\bar{y}=\bar{Y}+\bar{e}_y$ etc., the above estimator can be written as:

$$t_{50(1)}-\bar{Y}=\bar{e}_y-\alpha\bar{e}_x-\beta\bar{e}_z.$$

Squaring and applying expectation, we have:

$$MSE\left(t_{50(1)}\right)=E\left(t_{47(1)}-\bar{Y}\right)^2=E\left(\bar{e}_y-\alpha\bar{e}_x-\beta\bar{e}_z\right)^2. \quad (3.4.9)$$

Optimum values of α and β which minimizes (3.4.9) are:

$$\alpha=\frac{\bar{Y}C_y}{\bar{X}C_x}\left(\frac{\rho_{yx}-\rho_{yz}\rho_{xz}}{1-\rho_{xz}^2}\right)=\beta_{yx.z},$$

and

$$\beta=\frac{\bar{Y}C_y}{\bar{X}C_z}\left(\frac{\rho_{yz}-\rho_{yx}\rho_{xz}}{1-\rho_{xz}^2}\right)=\beta_{yz.x}$$

Using optimum values of α and β in (3.4.8) and simplifying, the mean square error of regression estimator for FIC is:

$$MSE\left(t_{50(1)}\right)=\theta_1\bar{Y}^2C_y^2\left(1-\rho_{y.xz}^2\right). \quad (3.4.10)$$

The mean square error of regression estimator for FIC, given in (3.4.10), is exactly the same as the mean square error of the Singh and Espejo (2003) estimator and generalized chain ratio estimator of Samiuddin and Hanif (2006). We can therefore say that these estimators are in equivalence class.

Comparing (3.4.2), (3.4.7) and (3.4.10) we can conclude that:

$$MSE\left(t_{50(1)}\right)<MSE\left(t_{49(1)}\right)<MSE\left(t_{48(1)}\right). \quad (3.4.11)$$

We can further conclude from inequality (3.4.11) that, whenever population information of auxiliary variables is utilized, then we will get more precise estimates.

3.4.2 Ratio Estimators

Samiuddin and Hanif (2006) have discussed different possible ratio estimators depending upon the availability of population information of the auxiliary variables. The first general ratio estimator considered by Samiuddin and Hanif (2006) is in the case when information of population mean of auxiliary variables X and Z is available. This estimator has been discussed in detail in section 3.3.2. The ratio estimator proposed by Samiuddin and Hanif (2006) when information on population mean of auxiliary variable Z is unavailable has the form:

$$t_{51(1)} = \bar{y}\left(\frac{\bar{X}}{\bar{x}}\right)^{\alpha} \bar{z}^{\beta}. \tag{3.4.12}$$

Another ratio estimator proposed by Samiuddin and Hanif (2006), when population information of both auxiliary variables is unknown is:

$$t_{52(1)} = \bar{y}\left(\bar{x}\right)^{\alpha}\left(\bar{z}\right)^{\beta}. \tag{3.4.13}$$

The mean square error of (3.4.12) and (3.4.13) are in complex form.

3.5 Some Shrinkage Estimators

The shrinkage method of estimation has been effectively used to develop more precise estimates of population characteristics. Searl (1964) proposed the shrinkage version of mean per unit estimator and shows that his proposed estimator is more precise as compared with classical mean per unit estimator. We have discussed the shrinkage mean per unit estimator in section 1.5. Samiuddin and Hanif (2006) have proposed the shrinkage version of various popular estimators including regression and ratio estimator and have shown that their proposed estimators are more precise. Shahbaz and Hanif (2009) have proposed a general method of obtaining a shrinkage version of an estimator for any population parameter alongside general form of mean square error of the shrinkage estimator. The general shrinkage estimator has been discussed in section 1.6. In this section we will present some shrinkage estimators of population mean proposed by Samiuddin and Hanif (2006). We will then present some applications of the general result of Shahbaz and Hanif (2009).

3.5.1 Shrinkage Regression Estimator

We have discussed the classical regression estimator in section 2.2.4. We have also seen that a number of estimators proposed by various survey statisticians are in equivalence class with this estimator. Samiuddin and Hanif (2006) proposed the following shrinkage version of classical regression estimator with single auxiliary variable:

$$t_{53(1)} = d_0\bar{y} + d_1\left(\bar{X} - \bar{x}\right), \tag{3.5.1}$$

where d_0 and d_1 are constants to be determined so that the mean square of (3.5.1) is minimized. Substituting $\bar{y}$ with $\bar{Y} + \bar{e}_y$ etc., we can write:

$$t_{53(1)} - \bar{Y} = d_0\bar{e}_y - d_1\bar{e}_x + \bar{Y}\left(d_0 - 1\right).$$

The mean square error of $t_{53(1)}$ is therefore:

$$MSE\left(t_{53(1)}\right)=E\left\{t_{53(1)}-\bar{Y}\right\}^2=E\left\{d_0\bar{e}_y-d_1\bar{e}_x+\bar{Y}\left(d_0-1\right)\right\}^2. \quad (3.5.2)$$

Differentiating (3.5.2) *w.r.t.* d_0 and d_1 and equating to zero, the optimum values of unknowns are $d_0=\left\{1+\theta_1 C_y^2\left(1-\rho_{xy}^2\right)\right\}^{-1}$ and $d_1=d_0\beta_{xy}$. Substituting optimum values of d_0 and d_1 in (3.5.2) and simplifying, the mean square error of $t_{53(1)}$ is:

$$MSE\left(t_{53(1)}\right)=\frac{\theta_1\bar{Y}^2C_y^2\left(1-\rho_{xy}^2\right)}{1+\theta_1C_y^2\left(1-\rho_{xy}^2\right)}. \quad (3.5.3)$$

The mean square error of $t_{53(1)}$, given in (3.5.3) is the same as the mean square error of the Lui (1990) estimator given in (3.2.11). The same mean square error can be readily obtained by using $\hat{\eta}=t_{4(1)}$ in (1.6.5).

Samiuddin and Hanif (2006) have proposed another shrinkage regression estimator with two auxiliary variables as:

$$t_{54(1)}=d_0\bar{y}+d_1\left(\bar{X}-\bar{x}\right)+d_2\left(\bar{Z}-\bar{z}\right). \quad (3.5.4)$$

Using $d_1=d_0\alpha$ and $d_2=d_0\beta$, the estimator (3.5.4) can be written as:

$$t_{54(1)}=d_0\left\{\bar{y}+\alpha\left(\bar{X}-\bar{x}\right)+\beta\left(\bar{Z}-\bar{z}\right)\right\}=d_0t_{48(1)}$$

The mean square error of $t_{54(1)}$ can be immediately written by using general result of Shahbaz and Hanif (2009) as:

$$MSE\left(t_{54(1)}\right)=\frac{\theta_1\bar{Y}^2C_y^2\left(1-\rho_{y.xz}^2\right)}{1+\theta_1C_y^2\left(1-\rho_{y.xz}^2\right)}. \quad (3.5.5)$$

Comparison of (3.5.5) with (3.4.3) immediately shows that $t_{54(1)}$ is more precise than $t_{48(1)}$. The general result of Shahbaz and Hanif (2009) can further be used to obtain a shrinkage version of regression estimators for partial information and no information cases suggested by Samiuddin and Hanif (2006).

3.5.2 Shrinkage Ratio and Ratio Type Estimators

The shrinkage ratio estimator has been proposed by Prasad (1989) and we have discussed that estimator in section 2.2.2. Samiuddin and Hanif (2006) proposed different versions of shrinkage ratio type estimators following the idea of Lui (1990). The shrinkage ratio type estimator proposed by Samiuddin and Hanif (2006) is given as:

$$t_{55(1)} = d_1\bar{y} + d_2\left(\frac{\bar{y}}{\bar{x}}\bar{X}\right) = d_1 t_{0(1)} + d_2 t_{1(1)} . \tag{3.5.6}$$

The estimator (3.5.6) is actually a weighted average of mean per unit and classical ratio estimator as discussed by Advaryu and Gupta (1983) but with the exception that sum of coefficients need not be 1. Using $\bar{y} = \bar{Y} + \bar{e}_y$ etc., the estimator (3.5.6) can be written as:

$$t_{55(1)} = \left(\bar{Y} + \bar{e}_y\right)\left[d_1 + d_2\left(1 + \frac{\bar{e}_x}{\bar{X}}\right)^{-1}\right].$$

Expanding $\left(1 + \bar{e}_x/\bar{X}\right)^{-1}$ and retaining the first order term only, we have:

$$t_{55(1)} = \left(\bar{Y} + \bar{e}_y\right)d_1 + d_2\left(\bar{Y} + \bar{e}_y - \frac{\bar{Y}}{\bar{X}}\bar{e}_x\right).$$

The mean square error of $t_{55(1)}$ is, therefore:

$$MSE\left(t_{55(1)}\right) = E\left[d_1\left(\bar{Y} + \bar{e}_y\right) + d_2\left(\bar{Y} + \bar{e}_y - \frac{\bar{Y}}{\bar{X}}\bar{e}_x\right) - \bar{Y}\right]^2 . \tag{3.5.7}$$

The optimum values of d_1 and d_2 which minimize (3.5.7) are given as:

$$d_1 = \frac{C_x - C_y\rho_{xy}}{C_x\left(1 + \theta C_y^2 - \theta C_y^2\rho_{xy}^2\right)} \text{ and } d_2 = \frac{C_y\rho_{xy}}{C_x\left(1 + \theta C_y^2 - \theta C_y^2\rho_{xy}^2\right)} .$$

Substituting the optimum values of d_1 and d_2 in (3.5.7) and simplifying, the mean square error of $t_{55(1)}$ is exactly the same as the mean square error of the Lui (1990) estimator and shrinkage regression estimator proposed by Samiuddin and Hanif (2006).

SECTION-II:

TWO-PHASE SAMPLING WITH AUXILIARY VARIABLE

CHAPTER FOUR

TWO-PHASE SAMPLING (ONE AND TWO AUXILIARY VARIABLES)

4.1 Introduction

In chapters 2 and 3 we have discussed the use of auxiliary variables in single phase sampling. Several situations may arise where we are interested in estimating the population mean of the variable of interest in the presence of an auxiliary variable, but sampling frame about the study variable and/or population information about auxiliary variable is not available. In such situations two-phase sampling may be used to decrease the costs of survey for a small loss of efficiency. Two-phase sampling has been an important topic under study since 1928. The presence or absence of population auxiliary information plays a vital role in the efficiency of two phase sampling estimators. Samiuddin and Hanif (2007) have discussed various possibilities of availability of population information of auxiliary variable(s).

i) Full Information Case.
ii) Partial Information Case.
iii) No Information Case

Samiuddin and Hanif (2007) have argued that single-phase sampling, if cost is not a matter of concern, is a suitable technique in the first situation. Samiuddin and Hanif (2007) have further argued that two phase sampling is a feasible method of estimation when population information of some or all auxiliary variables is not available.

In two-phase sampling a larger sample is selected from the population and information of the auxiliary variable is recorded, and at second phase information on the auxiliary variable and variable of interest is recorded for estimation. Two-phase sampling may, therefore, be used when the cost of drawing large sample is too high or when sampling frame and/or information of the auxiliary variable is not available for the population.

Two-phase sampling happens to be a powerful and cost effective procedure for finding reliable estimates in first-phase sample for the unknown parameters of the auxiliary variable X and hence has an eminent role to play in survey sampling [Hidiroglou and Sarndal (1995, 1998)]. Neyman (1938) has commented about two phase sampling as below:

> *"A more accurate estimate of the original character may be obtained for the same total expenditure by arranging the sampling of population in two steps. The first step is to secure data, for the second character only, from a relatively large random sample of the population in order to obtain an accurate estimate of the distribution of this character.*
>
> *The second step is to divide this sample, as in stratified sampling into classes or strata according to the value of the second character and to draw at random from each of the strata, a small sample for the costly intensive interviewing necessary to secure data regarding the first character.*
>
> *An estimate of the first character based on these samples may be more accurate than based on an equally expensive sample drawn at random without stratification. The question is to determine for a given expenditure, the sizes of the initial sample and the subsequent samples which yield the most accurate estimate of the first character"*

In an introduction to Neyman's paper, Dalenius (1954) stated:

> *"As demonstrated by the developments in the last half-century, supplementary information may be exploited for all aspects of the sample design, the definition of sampling units, the selection design and the estimation method."*

Rao (1973) has used two-phase sampling for stratification, non-response problems and analytic comparison. Sarndal and Swensson (1987) provided a framework for regression estimation in two-phase sampling. Chaudhuri and Roy (1994) studied the optimal properties of well-known regression estimators used in two-phase sampling. Breidt and Fuller (1993) gave a numerical estimation procedure for three-phase sampling, in the presence of auxiliary information.

In two-phase sampling, first phase sample of size n_1 is drawn from a population of size N and information on auxiliary variable(s) is recorded. A sub-sample of size n_2 ($n_2 < n_1$) is then drawn and information from the variable of interest and auxiliary variable(s) is obtained.

We now discuss popular estimators of two-phase sampling which are based upon information of a single auxiliary variable in the following section.

4.2 Ratio and Regression Estimators

In the following we discuss some popular ratio and regression estimators of population mean in two-phase sampling which are based upon information of a single auxiliary variable. We will discuss various possibilities of the estimators on the availability of auxiliary information.

4.2.1 Ratio Estimators

The ratio method of estimation has been very popular in single phase sampling. In two-phase sampling this method has also attracted a number of survey statisticians and a large number of estimators have been proposed from time to time. The ratio estimator, in two-phase sampling, may be constructed in two ways, viz. full and no information of the mean of the auxiliary variable. The ratio estimator when the mean of the auxiliary variable is known is given as:

$$t_{1(2)} = \frac{\bar{y}_2}{\bar{x}_2}\bar{X}. \tag{4.2.1}$$

The mean square error of (4.2.1) may immediately be written as:

$$MSE\left(t_{1(2)}\right) = \theta_2\bar{Y}^2\left(C_y^2 + C_x^2 - 2\rho_{xy}C_xC_y\right). \tag{4.2.2}$$

The mean square error given in (4.2.2) is more precise than the mean square error of the classical ratio estimator of single-phase sampling based upon a sample of size n_2.

Another ratio estimator of two-phase sampling when population information is not known is:

$$t_{2(2)} = \frac{\bar{y}_2}{\bar{x}_2}\bar{x}_1. \tag{4.2.3}$$

This estimator has much wider applicability as compared with (4.2.1) as this estimator does not require any population information except its size. The estimator (4.2.3) may be written by using section 1.4

$$t_{2(2)} = \frac{\bar{Y} + \bar{e}_{y_2}}{\bar{X} + \bar{e}_{x_2}}\left(\bar{X} + \bar{e}_{x_1}\right) = \left(\bar{Y} + \bar{e}_{y_2}\right)\left(1 + \bar{e}_{x_1}/\bar{X}\right)\left(1 + \bar{e}_{x_2}/\bar{X}\right)^{-1},$$

where $\bar{e}_{y_2}$ etc. are discussed in section 1.4. Expanding and retaining linear terms only we will have:

$$t_{2(2)} - \bar{Y} = \bar{e}_{y_2} + \frac{\bar{Y}}{\bar{X}}\left(\bar{e}_{x_1} - \bar{e}_{x_2}\right).$$

Squaring and applying expectations, the mean square error of $t_{2(2)}$ is:

$$MSE\left(t_{2(2)}\right)=E\left[\bar{e}_{y_2}+\frac{\bar{Y}}{\bar{X}}\left(\bar{e}_{x_1}-\bar{e}_{x_2}\right)\right]^2.$$

Expanding, using results from (1.4.1) and simplifying, the mean square error of $t_{2(2)}$ is:

$$MSE\left(t_{2(2)}\right)=\bar{Y}^2\left[\theta_2 C_y^2+\left(\theta_2-\theta_1\right)\left(C_x^2-2C_x C_y \rho_{xy}\right)\right], \quad (4.2.4)$$

or $$MSE\left(t_{2(2)}\right)=\theta_2\bar{Y}^2\left(C_y^2+C_x^2-2C_x C_y \rho_{xy}\right)-\theta_1\left(C_x^2-2C_x C_y \rho_{xy}\right),$$

or $$MSE\left(t_{2(2)}\right)=MSE\left(t_{1(1)}\right)-\theta_1\left(C_x^2-2C_x C_y \rho_{xy}\right). \quad (4.2.5)$$

Through comparison of (4.2.5) with (4.2.2) we see that $t_{2(2)}$ will be more precise than $t_{1(2)}$ if $\rho_{xy}>C_x/\left(2C_y\right)$. This is the same condition which we obtain in comparing classical ratio estimator in single-phase sampling with mean per unit estimator.

4.2.2 Sahoo and Sahoo (1994) Ratio Estimators

Sahoo and Sahoo (1994) proposed two ratio estimators for population mean when population information of the auxiliary variable is known. The first estimator proposed by Sahoo and Sahoo (1994) is:

$$t_{3(2)}=\frac{\bar{y}_2}{\bar{z}_2}\bar{Z}. \quad (4.2.6)$$

This is precisely the estimator $t_{1(2)}$ based upon auxiliary variable Z. The mean square error is the same as given in (4.2.2). Sahoo and Sahoo (1991) proposed another estimator by using information on both first and second phase sample alongside population information of the auxiliary variable. The estimator is:

$$t_{4(2)}=\frac{\bar{y}_2}{\bar{z}_1}\bar{Z}. \quad (4.2.7)$$

Substituting notations from section 1.4 in (4.2.7) and expanding up to linear term only we have:

$$t_{4(2)} - \bar{Y} = \bar{e}_{y_2} - \frac{\bar{Y}}{\bar{Z}}\bar{e}_{z_1}$$

The mean square error of $t_{4(2)}$ is, therefore:

$$MSE\left(t_{4(2)}\right) = E\left(t_{4(2)} - \bar{Y}\right)^2 = E\left(\bar{e}_{y_2} - \frac{\bar{Y}}{\bar{Z}}\bar{e}_{z_1}\right)^2.$$

Expanding the square, applying expectation and using (1.4.1), the mean square error of $t_{4(2)}$ is:

$$MSE\left(t_{4(2)}\right) = \bar{Y}^2\left[\theta_2 C_y^2 + \theta_1\left(C_z^2 - 2C_z C_y \rho_{yz}\right)\right]. \quad (4.2.8)$$

Comparison of (4.2.4) with (4.2.8) shows that estimator $t_{3(2)}$ will be more precise than $t_{4(2)}$ if $\rho > C_x/\left(2C_y\right)$ as notation for auxiliary variable is arbitrary.

4.2.3 Regression Estimators

We have seen in Chapter 2 and Chapter 3 that the regression method of estimation is more efficient than the ratio method of estimation. We have also seen that a number of estimators proposed by various survey statisticians is in equivalence class with the classical regression estimator. This method of estimation has attracted a large number of survey statisticians in two-phase sampling as well. The regression estimator in two-phase sampling may be constructed in two different ways; namely unknown and known information about the population parameter of the auxiliary variable. The regression estimator, for no information case is:

$$t_{5(2)} = \bar{y}_2 + \beta_{yx}\left(\bar{x}_1 - \bar{x}_2\right). \quad (4.2.9)$$

Substituting notations from section 1.4, we have:

$$t_{5(2)} - \bar{Y} = \bar{e}_{y_2} + \beta_{yx}\left(\bar{e}_{x_1} - \bar{e}_{x_2}\right).$$

Squaring and taking expectation of above equation, the mean square error of $t_{5(2)}$ will be:

$$MSE\left(t_{5(2)}\right) \approx E\left[\bar{e}_{y_2} + \beta_{yx}\left(\bar{e}_{x_1} - \bar{e}_{x_2}\right)\right]^2,$$

or $$MSE\left(t_{5(2)}\right)=E\left[\bar{e}_{y_2}^2+\beta_{yx}^2\left(\bar{e}_{x_1}-\bar{e}_{x_2}\right)^2+2\beta_{yx}\bar{e}_{y_2}\left(\bar{e}_{x_1}-\bar{e}_{x_2}\right)\right]. \quad (4.2.10)$$

Applying expectation on (4.2.10), using results from (1.4.1), putting the value of β_{yx} and simplifying, the mean square error of $t_{5(2)}$ will be:

$$MSE\left(t_{5(2)}\right)\approx\bar{Y}^2C_y^2\left[\theta_2+\left(\theta_1-\theta_2\right)\rho_{xy}^2\right],$$

or $$MSE\left(t_{5(2)}\right)\approx\bar{Y}^2C_y^2\left[\theta_2\left(1-\rho_{xy}^2\right)+\theta_1\rho_{xy}^2\right]. \quad (4.2.11)$$

Regression estimator proposed by Sahoo and Sahoo (1994) when population information of auxiliary variable is available for first phase. The estimator is:

$$t_{6(2)}=\bar{y}_2+\beta_{yx}\left(\bar{X}-\bar{x}_1\right);. \quad (4.2.12)$$

Using notations from section 1.4, we may write:

$$t_{6(2)}-\bar{Y}=\bar{e}_{y_2}-\beta_{yx}\bar{e}_{x_1}.$$

Squaring, applying expectation and using section 1.4, the mean square error of $t_{6(2)}$ is:

$$MSE\left(t_{6(2)}\right)=\bar{Y}^2C_y^2\left(\theta_2-\theta_1\rho_{yx}^2\right). \quad (4.2.13)$$

Comparison of (4.2.11) with (4.2.13) shows that $t_{5(2)}$ will be more precise than $t_{6(2)}$ if $\theta_2>2\theta_1$.

Another regression estimator when the population mean of the auxiliary variable is known is given as:

$$t_{7(2)}=\bar{y}_2+\beta_{yx}\left(\bar{X}-\bar{x}_2\right). \quad (4.2.14)$$

This is precisely classical regression estimator based on second phase sample. The mean square error can be immediately written parallel to (2.2.12).

4.2.4 Empirical Study

We have discussed seven different estimators in previous subsections. An empirical study has been conducted by Ahmad et al. (2007), Tahir (2008), and Bano (2009) independently by taking a different set of the population to decide about best estimator among the seven. The empirical study has been carried out by using 16 natural populations. The study has been conducted following three situations:

Case 1: All information from population is used.

Case 2: Some information from first phase and some information from second phase are used (means and coefficient of variation from first phase and coefficient of correlations from second phase).

Case 3: All information from second phase sample is used.

They conducted the empirical study by first computing mean square errors of seven discussed estimators and then ranking those mean square errors. The ranked mean square errors are given in tables A.1 to A.3. The results in the tables immediately show that $t_{7(2)}$ is best estimator in all three cases. The estimator $t_{2(2)}$ is second best in case-1 and $t_{4(2)}$ is second best in case-2.

4.3 Mohanty (1967) Estimator with Some Modifications

Mohanty (1967) proposed three estimators for population mean in two-phase sampling using information of two auxiliary variables. The proposed estimators are in line with the Mohanty (1967) estimator for single-phase sampling given in (2.3.1). Khare and Srivastava (1981) and Samiuddin and Hanif (2007) proposed certain modifications of Mohanty's (1967) estimator. First we discuss Mohanty's (1967) estimator along with its modification and second we will discuss Kare and Srivastava (1981) and Ahmed et al.'s (2007) modifications.

4.3.1 Mohanty's (1967) Regression-cum-Ratio Estimators

Mohanty (1967) proposed three estimators of population mean using information of two auxiliary variables. The first estimator for partial information is:

$$t_{8(2)} = \left[\bar{y}_2 + \beta_{yx}\left(\bar{x}_1 - \bar{x}_2\right)\right]\frac{\bar{Z}}{\bar{z}_2}. \tag{4.3.1}$$

Using $\bar{y}_2 = \bar{Y} + \bar{e}_{y_2}$ etc. in (4.3.1) and simplifying we have:

$$t_{8(2)} - \bar{Y} = \bar{e}_{y_2} + \beta_{yx}\left(\bar{e}_{x_1} - \bar{e}_{x_2}\right) - \frac{\bar{Y}}{\bar{Z}}\bar{e}_{z_2}.$$

The mean square error of (4.3.1) is:

$$MSE\left(t_{8(2)}\right)=E\left(t_{8(2)}-\bar{Y}\right)^2=E\left[\bar{e}_{y_2}+\beta_{yx}\left(\bar{e}_{x_1}-\bar{e}_{x_2}\right)-\frac{\bar{Y}}{\bar{Z}}\bar{e}_{z_2}\right]^2. \quad (4.3.2)$$

Expanding (4.3.2), putting the value of β_{yx} and using section 1.4, the mean square error of $t_{8(2)}$ may be written as:

$$MSE\left(t_{8(2)}\right)\approx\bar{Y}^2\left[\theta_2\left\{C_y^2\left(1-\rho_{yz}^2\right)+\left(C_z-C_y\rho_{yz}\right)^2\right\}\right.$$
$$\left.+\left(\theta_2-\theta_1\right)\left\{C_z^2\rho_{xz}^2-\left(C_y\rho_{xy}-C_z\rho_{xz}\right)^2\right\}\right]. \quad (4.3.3)$$

Mohanty (1967) proposed a second estimator based on no information:

$$t_{9(2)}=\left[\bar{y}_2+\beta_{yx}\left(\bar{x}_1-\bar{x}_2\right)\right]\frac{\bar{z}_1}{\bar{z}_2}. \quad (4.3.4)$$

Using notations from section 1.4, estimator $t_{9(2)}$ can be written as:

$$t_{9(2)}=\left[\left(\bar{Y}+\bar{e}_{y_2}\right)+\beta_{yx}\left(\bar{e}_{x_1}-\bar{e}_{x_2}\right)\right]\left(\frac{\bar{Z}+\bar{e}_{z_1}}{\bar{Z}+\bar{e}_{z_2}}\right).$$

Expanding and retaining linear terms only, we have:

$$t_{9(2)}-\bar{Y}=\bar{e}_{y_2}+\beta_{yx}\left(\bar{e}_{x_1}-\bar{e}_{x_2}\right)+\frac{\bar{Y}}{\bar{Z}}\left(\bar{e}_{z_1}-\bar{e}_{z_2}\right).$$

The mean square error of $t_{9(2)}$ is, therefore:

$$MSE\left(t_{9(2)}\right)\approx E\left[\bar{e}_{y_2}+\beta_{yx}\left(\bar{e}_{x_1}-\bar{e}_{x_2}\right)+\frac{\bar{Y}}{\bar{Z}}\left(\bar{e}_{z_1}-\bar{e}_{z_2}\right)\right]^2. \quad (4.3.5)$$

Substituting the value of β_{yx} in (4.3.5), expanding square and using section 1.4., the mean square error of $t_{9(2)}$ is:

$$MSE\left(t_{9(2)}\right)\approx\bar{Y}^2\left[\theta_2 C_y^2+\left(\theta_2-\theta_1\right)\left\{\rho_{xz}^2 C_z^2-\left(\rho_{xy}C_y-\rho_{xz}C_z\right)^2\right.\right.$$
$$\left.\left.+\left(C_z-C_y\rho_{yz}\right)^2-C_y^2\rho_{yz}^2\right\}\right]. \quad (4.3.6)$$

The estimator $t_{8(2)}$ has been modified by Samiuddin and Hanif (2007) by interchanging the role of two auxiliary variables. The modified estimator is:

$$t_{10(2)} = \left[\bar{y}_2 + \beta_{yz}\left(\bar{z}_1 - \bar{z}_2\right)\right]\frac{\bar{X}}{\bar{x}_2}. \tag{4.3.7}$$

The mean square error of $t_{10(2)}$ may be written immediately from (4.3.3) by interchanging subscript x with z as:

$$MSE\left(t_{10(2)}\right) \approx \bar{Y}^2\left[\theta_2\left\{C_y^2\left(1-\rho_{xy}^2\right)+\left(C_x - C_y\,\rho_{xy}\right)^2\right\}\right.$$
$$\left.+\left(\theta_2-\theta_1\right)\left\{C_x^2\,\rho_{xz}^2 - \left(C_y\rho_{yz} - C_x\rho_{xz}\right)^2\right\}\right]. \tag{4.3.8}$$

Mohanty (1967) also proposed third ratio-cum-regression type estimator when information about population mean of auxiliary variable X is available. The proposed estimator based on partial information is:

$$t_{11(2)} = \frac{\bar{y}_2}{\bar{z}_2}\left[\bar{z}_1 + \beta_{yx}\left(\bar{X} - \bar{x}_1\right)\right]. \tag{4.3.9}$$

Using notations from section 1.4, we may write:

$$t_{11(2)} - \bar{Y} = \bar{e}_{y_2} + \frac{\bar{Y}}{\bar{Z}}\left(\bar{e}_{z_1} - \bar{e}_{z_2}\right) - \frac{\bar{Y}}{\bar{Z}}\beta_{yx}\,\bar{e}_{x_1}.$$

Squaring, applying expectation and using value of β_{yx}, the mean square error of (4.3.9) may be written as:

$$MSE\left(t_{11(2)}\right) \approx \bar{Y}^2\left[\theta_2 C_y^2 + \left(\theta_2 - \theta_1\right)\left\{\left(C_z - C_y\rho_{yz}\right)^2 - C_y^2\rho_{yz}^2\right\}\right.$$
$$\left.+\theta_1\left\{\left(\bar{Y}C_y\rho_{xy}/\bar{Z} - C_y\rho_{xy}\right)^2 - C_y^2\rho_{xy}^2\right\}\right]. \tag{4.3.10}$$

Some more ratio type estimators for two-phase sampling have been proposed by Khare and Srivastava (1981) and Ahmad et al. (2007) when population information of an auxiliary variable is known. We now discuss those estimators in the following.

4.3.2 Khare and Srivastava (1981) Estimator with Some Modifications

Khare and Srivastava (1981) have proposed a regression-cum-ratio estimator when the population mean of one auxiliary variable is known in the regression part. The estimator is:

$$t_{12(2)} = \left[\bar{y}_2 + \beta_{yx}\left(\bar{X} - \bar{x}_1\right)\right]\frac{\bar{Z}}{\bar{z}_2}. \tag{4.3.11}$$

Using section 1.4, the estimator $t_{12(2)}$ may be written as:

$$t_{12(2)} = \left(\bar{Y} + \bar{e}_{y_2} - \beta_{yx}\bar{e}_{x_1}\right)\left(1 - \frac{\bar{e}_{z_2}}{\bar{Z}}\right),$$

or $\quad t_{12(2)} - \bar{Y} = \bar{e}_{y_2} - \beta_{yx}\bar{e}_{x_1} - \dfrac{\bar{Y}\bar{e}_{z_2}}{\bar{Z}}.$

The mean square error of $t_{12(2)}$ is:

$$MSE\left(t_{12(2)}\right) \approx E\left(t_{12(2)} - \bar{Y}\right)^2 = E\left(\bar{e}_{y_2} - \beta_{yx}\bar{e}_{x_1} - \frac{\bar{Y}\bar{e}_{z_2}}{\bar{Z}}\right)^2. \tag{4.3.12}$$

Expanding (4.3.12) and using (1.4.1), the mean square error $t_{12(2)}$ may be written as:

$$MSE\left(t_{12(2)}\right) \approx \bar{Y}^2\left[\theta_2\left(C_y^2 + C_z^2 - 2C_yC_z\rho_{yz}\right) - \theta_1 C_y\rho_{xy}\left(C_y\rho_{xy} - 2C_z\rho_{xz}\right)\right]. \tag{4.3.13}$$

Ahmad et al. (2007) has proposed the following three estimators, following the lines of Mohanty (1967) and Khare and Srivastava (1981), when the population mean of one of the auxiliary variables is known, i.e. the Partial information case.

$$t_{13(2)} = \left[\bar{y}_2 + \beta_{xy}\left(\bar{X} - \bar{x}_1\right)\right]\frac{\bar{z}_2}{\bar{Z}}, \tag{4.3.14}$$

$$t_{14(2)} = \left[\bar{y}_2 + \beta_{xy}\left(\bar{X} - \bar{x}_2\right)\right]\frac{\bar{Z}}{\bar{z}_2}, \tag{4.3.15}$$

and

$$t_{15(2)} = \left[\bar{y}_2 + \beta_{xy}\left(\bar{X} - \bar{x}_2\right)\right]\frac{\bar{Z}}{\bar{z}_1}. \tag{4.3.16}$$

The estimator $t_{13(2)}$ is a product type estimator parallel to $t_{12(2)}$. The estimators $t_{14(2)}$ and $t_{15(2)}$ are regression-cum-ratio type estimators with a changed role of auxiliary information. The mean square errors of the above three estimators may be easily derived, and are:

$$MSE\left(t_{13(2)}\right) \approx \bar{Y}^2\left[\theta_2\left(C_y^2 + C_z^2 + 2C_yC_z\rho_{yz}\right) - \theta_1 C_y\rho_{xy}\left(C_y\rho_{xy} + 2C_z\rho_{zx}\right)\right], \tag{4.3.17}$$

$$MSE\left(t_{14(2)}\right) \approx \bar{Y}^2\theta_2\left[C_y^2\left(1-\rho_{xy}^2\right) + C_z^2 - 2C_yC_z\left(\rho_{yz} - \rho_{xy}\rho_{zx}\right)\right] \tag{4.3.18}$$

and

$$MSE\left(t_{15(2)}\right) \approx \bar{Y}^2\left[\theta_2 C_y^2\left(1-\rho_{xy}^2\right) + \theta_1 C_z^2 - 2\theta_1 C_yC_z\left(\rho_{yz} - \rho_{xy}\rho_{zx}\right)\right]. \tag{4.3.19}$$

We have conducted the empirical study to see the performance of $t_{8(2)}$ to $t_{15(2)}$ in the following subsection.

4.3.3 Empirical Study

The direct comparison of these estimators is not easy. Therefore, an empirical study has been conducted by Ahmad et al. (2007), Tahir (2008) and Bano (2009) to decide the best estimator in the class of the Mohanty estimator and its modified versions. The empirical study has been carried out by using sixteen natural populations on the pattern discussed in section 3.2.4. The mean square error of the Mohanty (1967) estimator and its modified versions are computed. After computing the mean square errors ranking has been done. The results of these rankings are given in Tables A.4 through A.6. A close look at these tables immediately shows that estimator $t_{15(2)}$ is best among this class for all possibilities and is followed by $t_{12(2)}$. The estimator $t_{13(2)}$ is worst in this class.

4.4 Chain Based Ratio-Type Estimator and its Modifications

We have discussed the regression-cum-ratio type estimator proposed by Mohanty (1967) and its modifications by Khare and Srivastava (1981), Samiuddin and Hanif (2006), and Ahmad et al. (2007) which are based upon information of two auxiliary variables. Chand (1975), following

Swain (1970), proposed his popular chain ratio-type estimator in two-phase sampling based upon two auxiliary variables. In this section we discuss the estimator proposed by Chand (1975) along with its modification.

4.4.1 Chand's (1975) Chain-Based Ratio-Type Estimator

Chand (1975) proposed the following chain ratio type estimator based upon information of two auxiliary variables:

$$t_{16(2)} = \bar{y}_2 \frac{\bar{x}_1}{\bar{x}_2} \frac{\bar{Z}}{\bar{z}_1}. \tag{4.4.1}$$

The estimator $t_{16(2)}$ uses information at both phases along with population information of one auxiliary variable. Using section 1.4, we have, from (4.4.1):

$$t_{16(2)} - \bar{Y} = \bar{e}_{y_2} + \bar{Y}\left(\bar{e}_{x_1} - \bar{e}_{x_2}\right)/\bar{X} - \bar{Y}\bar{e}_{z_1}/\bar{Z}. \tag{4.4.2}$$

Squaring (4.4.2) and using (1.4.1), the mean square error of $t_{16(2)}$ is:

$$MSE\left(t_{16(2)}\right) \approx \bar{Y}^2\Big[\theta_2 C_y^2 + \left(\theta_2 - \theta_1\right)C_x^2 + \theta_1 C_z^2$$
$$+ 2\left(\theta_1 - \theta_2\right)C_y C_x \rho_{xy} - 2\theta_1 C_y C_z \rho_{yz}\Big].$$

Simplifying the above expression we have:

$$MSE\left(t_{16(2)}\right) \approx \bar{Y}^2\Big[\theta_2 C_y^2 + \left(\theta_2 - \theta_1\right)\left\{\left(C_x - C_y\rho_{xy}\right)^2 - C_y^2\rho_{xy}^2\right\}$$
$$+\theta_1\left\{\left(C_z - C_y\rho_{yz}\right)^2 - C_y^2\rho_{yz}^2\right\}\Big]. \tag{4.4.3}$$

A modification of (4.4.1) has been proposed by Samiuddin and Hanif (2006) by interchanging the role of auxiliary variables, i.e.:

$$t_{17(2)} = \bar{y}_2 \frac{\bar{z}_1}{\bar{z}_2} \frac{\bar{X}}{\bar{x}_1}. \tag{4.4.4}$$

The mean square error of $t_{17(2)}$ may immediately be written parallel to (4.4.3) just by interchanging subscript x with z.

4.4.2 Sahoo and Sahoo (1993) Chain-based Ratio-Type Estimators

Sahoo and Sahoo (1993) following Chand (1975) proposed two chain-based estimators when population information of one auxiliary variable is known. The first of two estimators proposed by Sahoo and Sahoo (1993) is a chain product estimator:

$$t_{18(2)} = \bar{y}_2 \frac{\bar{x}_2}{\bar{x}_1} \frac{\bar{z}_1}{\bar{\bar{Z}}} . \tag{4.4.5}$$

The estimator $t_{18(2)}$ uses information at both phases along with population information of one auxiliary variable. Using notation from section 1.4 we can write (4.4.5) as:

$$t_{18(2)} - \bar{Y} = \bar{e}_{y_2} - \frac{\bar{Y}}{\bar{\bar{X}}}\left(\bar{e}_{x_1} - \bar{e}_{x_2}\right) + \frac{\bar{Y}}{\bar{\bar{Z}}}\bar{e}_{z_1} . \tag{4.4.6}$$

Squaring (4.4.6), applying expectation and using section 1.4, the mean square error of $t_{18(2)}$ is:

$$MSE\left(t_{18(2)}\right) \approx \bar{Y}^2 \Big[\theta_2 C_y^2 + \left(\theta_2 - \theta_1\right)\left\{\left(C_x + \rho_{xy} C_y\right)^2 - C_y^2 \rho_{xy}^2\right\}$$
$$+ \theta_1 \left\{\left(C_z + \rho_{yz} C_y\right)^2 - \rho_{yz}^2 C_y^2\right\}\Big] . \tag{4.4.7}$$

The difference in mean square error of $t_{16(2)}$ and $t_{18(2)}$ is the same as the difference between the mean square error of the classical ratio and product estimators. Sahoo and Sahoo (1993) have proposed another modification of Chand's (1975) estimator. The proposed modified estimator is:

$$t_{19(2)} = \bar{y}_2 \frac{\bar{x}_2}{\bar{x}_1} \frac{\bar{Z}}{\bar{z}_1} . \tag{4.4.8}$$

Substituting $\bar{y}_2 = \bar{Y} + \bar{e}_{y_2}$ etc. in (4.4.8) we have:

$$t_{19(2)} - \bar{Y} = \bar{e}_{y_2} - \frac{\bar{Y}}{\bar{\bar{X}}}\left(\bar{e}_{x_1} - \bar{e}_{x_2}\right) - \frac{\bar{Y}}{\bar{\bar{Z}}}\bar{e}_{z_1} .$$

The mean square error of $t_{19(2)}$ is given as:

$$MSE\left(t_{19(2)}\right) \approx E\left[\bar{e}_{y_2} - \frac{\bar{Y}}{\bar{X}}\left(\bar{e}_{x_1} - \bar{e}_{x_2}\right) - \frac{\bar{Y}}{\bar{Z}}\bar{e}_{z_1}\right]^2. \quad (4.4.9)$$

Expanding (4.4.9), applying expectation and using section 1.4 the mean square error of $t_{19(2)}$ may be written as:

$$MSE\left(t_{19(2)}\right) \approx \bar{Y}^2\left[\theta_1\left(C_y^2 + C_z^2 - 2C_yC_z\rho_{yz}\right) + \left(\theta_2 - \theta_1\right)\left(C_y^2 + C_x^2 + 2C_yC_x\rho_{xy}\right)\right]. \quad (4.4.10)$$

The mean square error (4.4.10) is actually a weighted sum of two mean square errors, one of classical ratio estimator and the other of classical product estimators.

Sahoo and Sahoo (1993) have proposed a third chain-based estimator which is a ratio-cum-product estimator. The estimator is:

$$t_{20(2)} = \bar{y}_2 \frac{\bar{x}_1}{\bar{x}_2}\frac{\bar{z}_1}{\bar{Z}}. \quad (4.4.11)$$

The mean square error of $t_{20(2)}$ may be easily written from (4.4.10).

Samiuddin and Hanif (2006) have proposed some further modifications of the Sahoo and Sahoo (1993) estimators. We discuss these modifications in the following subsection.

4.4.3 Modification of Sahoo and Sahoo (1993) Chain-based Estimator

Samiuddin and Hanif (2006) have proposed several modifications of the Chand (1975) and Sahoo and Sahoo (1993) estimators by interchanging the role of variables. These estimators are parallel to single phase estimators of Samiuddin and Hanif (2006) given in (2.4.1). The chain ratio type estimators proposed by Samiuddin and Hanif (2006) are:

$$t_{21(2)} = \bar{y}_2\left(\bar{z}_2/\bar{z}_1\right)\left(\bar{X}/\bar{x}_1\right), \quad (4.4.12)$$

$$t_{22(2)} = \bar{y}_2\left(\bar{z}_1/\bar{z}_2\right)\left(\bar{x}_1/\bar{X}\right), \quad (4.4.13)$$

$$t_{23(2)} = \bar{y}_2\left(\bar{x}_1/\bar{X}\right)\left(\bar{z}_1/\bar{Z}\right), \quad (4.4.14)$$

$$t_{24(2)} = \bar{y}_2\left(\bar{z}_1/\bar{z}_2\right)\left(\bar{x}_2/\bar{x}_1\right), \quad (4.4.15)$$

$$t_{25(2)} = \bar{y}_2\left(\bar{X}/\bar{x}_2\right)\left(\bar{Z}/\bar{z}_2\right), \quad (4.4.16)$$

$$t_{26(2)} = \bar{y}_2\left(\bar{X}/\bar{x}_1\right)\left(\bar{Z}/\bar{z}_2\right), \quad (4.4.17)$$

$$t_{27(2)} = \bar{y}_2\left(\bar{X}/\bar{x}_2\right)\left(\bar{Z}/\bar{z}_1\right), \quad (4.4.18)$$

and

$$t_{28(2)} = \bar{y}_2\left(\bar{X}/\bar{x}_1\right)\left(\bar{Z}/\bar{z}_1\right). \quad (4.4.19)$$

The estimator $t_{24(2)}$ is the only estimator which is based upon complete sample information. Samiuddin and Hanif (2006) have argued that the estimators $t_{25(2)}$ to $t_{28(2)}$ use population information of both auxiliary variables. The mean square errors of modified estimators are:

$$MSE\left(t_{21(2)}\right) \approx \bar{Y}^2\left[\theta_1\left(C_y^2 + C_x^2 - 2C_yC_x\rho_{xy}\right) + \left(\theta_2 - \theta_1\right)\left(C_y^2 + C_z^2 + 2C_yC_z\rho_{yz}\right)\right], \quad (4.4.20)$$

$$MSE\left(t_{22(2)}\right) \approx \bar{Y}^2\left[\theta_2C_y^2 + \left(\theta_2 - \theta_1\right)\left\{\left(C_z - C_y\rho_{yz}\right)^2 - C_y^2\rho_{yz}^2\right\} + \theta_1\left\{\left(C_x + C_y\,\rho_{xy}\right)^2 - C_y^2\rho_{xy}^2\right\}\right], \quad (4.4.21)$$

$$MSE\left(t_{23(2)}\right) \approx \bar{Y}^2\left[\theta_1\left\{C_x^2 + 2C_x\left(C_y\rho_{xy} + C_z\rho_{xz}\right)\right\}, + \theta_2\left\{C_y^2 + C_z\left(C_z + 2C_y\rho_{yz}\right)\right\}\right] \quad (4.4.22)$$

$$MSE\left(t_{24(2)}\right) \approx \bar{Y}^2\left[\theta_2C_y^2 + \left(\theta_2 - \theta_1\right)\left\{C_x^2 + C_z^2, + 2\left(C_yC_x\rho_{xy} - C_yC_z\rho_{yz} - C_xC_z\rho_{xz}\right)\right\}\right] \quad (4.4.23)$$

$$MSE\left(t_{25(2)}\right) \approx \theta_2\bar{Y}^2\left[C_y^2 + C_x^2 + C_z^2 - 2C_yC_x\rho_{xy}, - 2C_yC_z\rho_{yz} + 2C_xC_z\rho_{xz}\right] \quad (4.4.24)$$

$$MSE\left(t_{26(2)}\right) \approx \bar{Y}^2\left[\theta_2\left(C_y^2 + C_z^2 - 2C_yC_z\rho_{yz}\right) + \theta_1\left(C_x^2 - 2C_yC_x\rho_{xy} + 2C_xC_z\rho_{xz}\right)\right], \quad (4.4.25)$$

$$MSE\left(t_{27(2)}\right) \approx \bar{Y}^2\left[\theta_2\left(C_y^2 + C_x^2 - 2C_yC_x\rho_{xy}\right) + \theta_1\left(C_z^2 - 2C_yC_z\rho_{yz} + 2C_xC_z\rho_{xz}\right)\right], \quad (4.4.26)$$

and

$$MSE\left(t_{28(2)}\right)\approx\bar{Y}^2\left[\theta_2C_y^2+\theta_1\left(C_x^2+C_z^2-2C_yC_x\rho_{xy}\right.\right.$$
$$\left.\left.-2C_yC_z\rho_{yz}+2C_xC_z\rho_{xz}\right)\right]. \tag{4.4.27}$$

Ahmad et al. (2007) have conducted an empirical study to choose the best estimator among the chain-based ratio estimators.

4.4.4 Empirical Study

We have discussed thirteen chain-based estimators proposed by Chand (1975), Sahoo and Sahoo (1993), Samiuddin and Hanif (2006) and Ahmad et al. (2007). Ahmad et al. (2007) conducted an empirical study on the pattern discussed in section 4.3.3. The ranked mean square errors of thirteen estimators are given in Table A.7 to A.9. Looking at the tables we can readily conclude that the chain estimator $t_{16(2)}$ proposed by Chand (1975) is best among this class and is followed by $t_{17(2)}$ proposed by Samiuddin and Hanif (2006).

4.5 Kiregyera's (1980, 84) Estimators and Some Modifications

Kiregyera (1980, 1984) used the idea of Mohanty (1967) to suggest some more estimators for two phase sampling using information of two auxiliary variables. We now discuss estimators proposed by Kiregyera (1980, 1984) and their modifications suggested by Samiuddin and Hanif (2006).

4.5.1 Kiregyera's (1980) Ratio-cum-Regression Estimator

Kiregyera (1980) suggested an estimator on the pattern of Mohanty's third estimator, $t_{11(2)}$ using information of two auxiliary variables. The proposed estimator is a regression-in-ratio estimator. The estimator proposed by Kiregyera (1980) differs from Mohanty (1967) estimator in that the latter estimator regression weight of the estimand on auxiliary regression is used, whereas in the former regression estimator of one auxiliary variable on the other auxiliary variable is used. The estimator proposed by Kiregyera (1980) is:

$$t_{29(2)} = \frac{\bar{y}_2}{\bar{x}_2}\left[\bar{x}_1 + \beta_{xz}\left(\bar{Z} - \bar{z}_1\right)\right]. \tag{4.5.1}$$

The mean square error of (4.5.1) may be derived by using section 1.4 and for this (4.5.1) may be written as:

$$t_{29(2)} - \bar{Y} = \bar{e}_{y_2} + \frac{\bar{Y}}{\bar{X}}\left(\bar{e}_{x_1} - \bar{e}_{x_2}\right) - \beta_{xz}\frac{\bar{Y}}{\bar{X}}\bar{e}_{z_1}$$

Squaring and applying expectation, the mean square error of $t_{29(2)}$ is:

$$MSE\left(t_{29(2)}\right) \approx E\left[\bar{e}_{y_2} + \frac{\bar{Y}}{\bar{X}}\left(\bar{e}_{x_1} - \bar{e}_{x_2}\right) - \beta_{xz}\frac{\bar{Y}}{\bar{X}}\bar{e}_{z_1}\right]^2.$$

Expanding the square, using (1.4.1) and simplifying, the mean square error of $t_{29(2)}$ is:

$$MSE\left(t_{29(2)}\right) \approx \bar{Y}^2\left[\theta_2 C_y^2 + \left(\theta_2 - \theta_1\right)\left\{\left(C_x - \rho_{xy} C_y\right)^2 - \rho_{xy}^2 C_y^2\right\}\right.$$
$$\left. + \theta_1\left\{\left(C_x\rho_{xz} - C_y\rho_{yz}\right)^2 - C_y^2\rho_{yz}^2\right\}\right]. \tag{4.5.2}$$

Samiuddin and Hanif (2006) have suggested a modification of the Kiregyera (1984) estimator by interchanging the role of auxiliary variables. The estimator proposed by Samiuddin and Hanif (2006) is:

$$t_{30(2)} = \frac{\bar{y}_2}{\bar{z}_2}\left[\bar{z}_1 + b_{xz}\left(\bar{X} - \bar{x}_1\right)\right]. \tag{4.5.3}$$

The mean square error of (4.5.3) may be immediately written from (4.5.2) by interchanging subscripts x and z.

4.5.2 Kiregyera's (1984) Estimator

Kiregyera (1984) proposed the following ratio-in-regression estimator by replacing the mean of auxiliary variable X with its estimate of the first phase sample.

$$t_{31(2)} = \bar{y}_2 + \beta_{yx}\left(\frac{\bar{x}_1}{\bar{z}_1}\cdot\bar{Z} - \bar{x}_2\right). \tag{4.5.4}$$

Replacing $\bar{y}_2$ with $\bar{Y} + \bar{e}_{y_2}$ etc., we have from (4.5.4):

$$t_{31(2)} - \bar{Y} = \bar{e}_{y_2} + \beta_{yx}\left(\bar{e}_{x_1} - \bar{e}_{x_2}\right) - \beta_{yx}\frac{\bar{X}}{\bar{\bar{Z}}}\bar{e}_{z_1}.$$

The mean square error of $t_{31(2)}$ is:

$$MSE\left(t_{31(2)}\right) \approx E\left[\bar{e}_{y_2} + \beta_{yx}\left(\bar{e}_{x_1} - \bar{e}_{x_2}\right) - \beta_{yx}\frac{\bar{X}}{\bar{\bar{Z}}}\bar{e}_{z_1}\right]^2.$$

Expanding the R.H.S, using (1.4.1) and simplifying, the mean square error of $t_{31(2)}$ is given as:

$$MSE\left(t_{31(2)}\right) \approx \bar{Y}^2 C_y^2\left[\theta_2\left(1-\rho_{xy}^2\right) + \theta_1\left\{\left(\rho_{xy}C_z/C_x - \rho_{yz}\right)^2 - \rho_{yz}^2 + \rho_{xy}^2\right\}\right]. \tag{4.5.5}$$

A modification of $t_{31(2)}$, suggested by Samiuddin and Hanif (2006), is:

$$t_{32(2)} = \bar{y}_2 + \beta_{yz}\left(\frac{\bar{z}_1}{\bar{x}_1}\bar{X} - \bar{z}_2\right). \tag{4.5.6}$$

Again, the mean square error of (4.5.6) may be written from (4.5.5) by changing subscripts x and z.

4.5.3 Kiregyera's (1984) Regression- in- Regression Estimator

Kiregyera (1984) has proposed the following regression-in-regression estimator:

$$t_{33(2)} = \bar{y}_2 + \beta_{yx}\left\{\left(\bar{x}_1 - \bar{x}_2\right) + \beta_{xz}\left(\bar{Z} - \bar{z}_1\right)\right\}. \tag{4.5.7}$$

The estimator (4.5.7) is simpler as compared with the Kiregyera (1980, 1984) estimators given in (4.5.1) and (4.5.4). Using notations from section 1.4, we have:

$$t_{33(2)} - \bar{Y} = \bar{e}_{y_2} + \beta_{yx}\left(\bar{e}_{x_1} - \bar{e}_{x_2}\right) - \beta_{yx}\beta_{xz}\bar{e}_{z_1}.$$

Squaring and applying expectation, we have:

$$MSE\left(t_{32(2)}\right) \approx E\left[\bar{e}_{y_2} + \beta_{yx}\left(\bar{e}_{x_1} - \bar{e}_{x_2}\right) - \beta_{yx}\beta_{xz}\bar{e}_{z_1}\right]^2.$$

Simplifying and using (1.4.1), the mean square error of $t_{33(2)}$ is:

$$MSE\left(t_{33(2)}\right) \approx \bar{Y}^2 C_y^2 \left[\theta_2 \left(1-\rho_{xy}^2\right) + \theta_1 \left(\rho_{xy}^2 - \rho_{yz}^2\right) + \theta_1 \left(\rho_{yz} - \rho_{xy}\rho_{xz}\right)^2 \right]. \tag{4.5.8}$$

Samiuddin and Hanif (2006) have proposed the modification of (4.5.7) by interchanging the role of auxiliary variables X and Z.

4.5.4 Sahoo and Sahoo (1993) Modified Estimators

Sahoo and Sahoo (1993) used Kiregyera's (1980) idea to propose two estimators of population mean in two-phase sampling using two auxiliary variables. The first of two estimators proposed by Sahoo and Sahoo (1993) is given as:

$$t_{34(2)} = \bar{y}_2 \frac{\bar{x}_2}{\bar{x}_1 + \beta_{xz}\left(\bar{Z} - \bar{z}_1\right)}. \tag{4.5.9}$$

The mean square error of $t_{34(2)}$ may be derived by using $\bar{y}_2 = \bar{Y} + \bar{e}_{y_2}$ etc. in (4.5.9) and writing:

$$t_{34(2)} - \bar{Y} = \bar{e}_{y_2} - \frac{\bar{Y}}{\bar{X}}\left(\bar{e}_{x_1} - \bar{e}_{x_2}\right) + \beta_{xz}\frac{\bar{Y}}{\bar{X}}\bar{e}_{z_1}$$

Squaring, applying expectation and simplifying, the mean square error of $t_{34(2)}$ is:

$$MSE\left(t_{34(2)}\right) \approx \bar{Y}^2 \left[\theta_2 C_y^2 \left(1-\rho_{xy}^2\right) + \theta_1 C_y^2 \left(\rho_{xy}^2 - \rho_{yz}^2\right) + \left(\theta_2 - \theta_1\right)\left(C_x + C_y\rho_{xy}\right)^2 + \theta_1 \left(C_x\rho_{xz} + C_y\rho_{yz}\right)^2 \right]. \tag{4.5.10}$$

Sahoo and Sahoo (1993) in their second estimator used the idea suggested by Kiregyera (1984) proposed following modification of $t_{31(2)}$ by using the product estimator instead of the ratio estimator in (4.5.4). The modified estimator is:

$$t_{35(2)} = \bar{y}_2 + \beta_{yx}\left(\bar{x}_1 \frac{\bar{z}_1}{\bar{Z}} - \bar{x}_2\right). \tag{4.5.11}$$

The mean square of (4.5.11) may be written immediately as:

$$MSE\left(t_{35(2)}\right) \approx C_y^2\bar{Y}^2\left[\theta_2\left(1-\rho_{xy}^2\right)+\theta_1\left(\rho_{xy}^2-\rho_{yz}^2\right)+\theta_1\left(\rho_{xy}C_z/C_x+\rho_{yz}\right)^2\right]. \tag{4.5.12}$$

4.5.5 Empirical Study

Samiuddin and Hanif (2006) used sixteen natural populations for an empirical study to pick the best among the estimators discussed in section 4.5. The empirical study was carried out on the pattern of section 4.3 and 4.4. We have given the results of the empirical study in the form of ranked mean square error of various estimators under different sampling mechanisms in Tables A.10 to A.12. The results of empirical study clearly show that estimator $t_{33(2)}$ is the best estimator in this class in all cases, and is followed by $t_{29(2)}$. The estimator $t_{35(2)}$ is the worst estimator in this class.

4.6 Mukerjee-Rao-Vijayan (1987) Estimators

Kiregyera (1980, 1984) proposed several ratio-in-regression and regression-in-ratio estimators based upon information of two auxiliary variables. Mukerjee et al. (1987) proposed three regression type estimators. These are simpler to use as compared with the Kiregyera (1980, 1984) estimators. Mukerjee et al. (1987) used a simple regression method to propose estimators based upon information of two auxiliary variables. The first of three estimators proposed by Mukerjee et al. (1987) is based upon all sample information and is given as:

$$t_{36(2)} = \bar{y}_2+\beta_{yx}\left(\bar{x}_1-\bar{x}_2\right)+\beta_{yz}\left(\bar{z}_1-\bar{z}_2\right). \tag{4.6.1}$$

The estimator $t_{36(2)}$ is like multiple regression estimators based upon information of first and second phase with the difference that simple regression weights are used instead of partial regression weights. The mean square error of $t_{36(2)}$ may be derived by using $\bar{y}_2=\bar{Y}+\bar{e}_{y_2}$ etc. in (4.6.1) and writing:

$$t_{36(2)}-\bar{Y}=\bar{e}_{y_2}+\beta_{yx}\left(\bar{e}_{x_1}-\bar{e}_{x_2}\right)+\beta_{yz}\left(\bar{e}_{z_1}-\bar{e}_{z_2}\right). \tag{4.6.2}$$

The mean square error of $t_{36(2)}$ is, therefore:

$$MSE\left(t_{36(2)}\right)=E\left[\bar{e}_{y_2}+\beta_{yx}\left(\bar{e}_{x_1}-\bar{e}_{x_2}\right)+\beta_{yz}\left(\bar{e}_{z_1}-\bar{e}_{z_2}\right)\right]^2$$

Expanding the square and using (1.4.1), the mean square error of $t_{36(2)}$ is:

$$MSE\left(t_{36(2)}\right)\approx\bar{Y}^2C_y^2\left[\theta_2-(\theta_2-\theta_1)\left(\rho_{xy}^2+\rho_{yz}^2-2\rho_{xy}\rho_{yz}\rho_{xz}\right)\right],$$

or

$$MSE\left(t_{36(2)}\right)\approx\bar{Y}^2C_y^2\left[\theta_2\left\{1-\rho_{y.xz}^2\left(1-\rho_{xz}^2\right)\right\}+\theta_1\left(1-\rho_{xz}^2\right)\rho_{y.xz}^2\right]. \quad (4.6.3)$$

Mukerjee et al. (1987) proposed another their second estimator by using population information of one of the auxiliary variables. The estimator is:

$$t_{37(2)}=\bar{y}_2+\beta_{yx}\left(\bar{x}_1-\bar{x}_2\right)+\beta_{yz}\left(\bar{Z}-\bar{z}_2\right). \quad (4.6.4)$$

Using notations from section 1.4, we may write (4.6.4) as:

$$t_{37(2)}-\bar{Y}=\bar{e}_{y_2}+\beta_{yx}\left(\bar{e}_{x_1}-\bar{e}_{x_2}\right)-\beta_{yz}\left(\bar{e}_{z_2}\right).$$

Squaring, applying expectations and using 1.4.1, the mean square error of $t_{37(2)}$ can be written as:

$$MSE\left(t_{37(2)}\right)\approx\bar{Y}^2C_y^2\left[\theta_2-\theta_1\rho_{yz}^2-(\theta_2-\theta_1)\left(\rho_{xy}^2+\rho_{yz}^2-2\rho_{xy}\rho_{yz}\rho_{xz}\right)\right]$$

or

$$MSE\left(t_{37(2)}\right)\approx\bar{Y}^2C_y^2\left[\theta_2\left\{1-\rho_{y.xz}^2\left(1-\rho_{xz}^2\right)\right\}+\theta_1\left\{\rho_{y.xz}^2\left(1-\rho_{xz}^2\right)-\rho_{yz}^2\right\}\right]. \quad (4.6.5)$$

A comparison of (4.6.3) and (4.6.5) clearly shows that $MSE\left(t_{37(2)}\right)<MSE\left(t_{36(2)}\right)$ and hence estimator $t_{37(2)}$ is more precise as compared with $t_{36(2)}$.

Mukerjee et al. (1987) proposed their third regression-type estimator, i.e.:

$$t_{38(2)}=\bar{y}_2+b_{yx}\left(\bar{x}_1-\bar{x}_2\right)+b_{yx}b_{xz}\left(\bar{Z}-\bar{z}_1\right)+b_{yz}\left(\bar{Z}-\bar{z}_2\right). \quad (4.6.6)$$

The estimator $t_{38(2)}$ is different from $t_{37(2)}$ in that the former uses information at both phases along with population information of one auxiliary variable. Using notation from section 1.4, in (4.6.6) we have:

$$t_{38(2)} - \bar{Y} = \bar{e}_{y_2} + \beta_{yx}\left(\bar{e}_{x_1} - \bar{e}_{x_2}\right) - \beta_{yx}\beta_{xz}\bar{e}_{z_1} - \beta_{yz}\bar{e}_{z_2}.$$

The mean square error may immediately be written as:

$$MSE\left(t_{38(2)}\right) = \bar{Y}^2 C_y^2\left[\theta_1\left(\rho_{yz} - \rho_{xy}\rho_{yz}\right)^2 + \theta_2\left(1 - \rho_{yz}^2 - \rho_{xy}^2 + 2\rho_{xy}\rho_{yz}\rho_{xz}\right)\right]. \tag{4.6.7}$$

The mean square error given in (4.6.7) is larger as compared with mean square error given in (4.6.5).

4.7 Sahoo-Sahoo-Mohanty (1993) Estimators with Modifications

Following the lines of Mukerjee et al. (1987), Sahoo et al. (1993) also proposed some regression type estimators in two-phase sampling.

4.7.1 Sahoo-Sahoo-Mohanty (1993) Regression-Type Estimator-I

Sahoo et al. (1993) proposed the following regression type estimator which uses population information of one auxiliary variable.

$$t_{39(2)} = \bar{y}_2 + \beta_{yx}\left(\bar{x}_1 - \bar{x}_2\right) + \beta_{yz}\left(\bar{Z} - \bar{z}_1\right). \tag{4.7.1}$$

The estimator $t_{39(2)}$ is like estimator $t_{37(1)}$ with the difference that the former uses information of auxiliary variable Z at the first phase along with population information, whereas the latter uses information of auxiliary variable Z at the second phase alongside population information. Using notations from section 1.4 we may write:

$$t_{39(2)} - \bar{Y} = \bar{e}_{y_2} + \beta_{yx}\left(\bar{e}_{x_1} - \bar{e}_{x_2}\right) - \beta_{yz}\,\bar{e}_{z_1}.$$

The mean square error of $t_{39(2)}$ is, therefore:

$$MSE\left(t_{39(2)}\right) \approx E\left[\bar{e}_{y_2} + \beta_{yx}\left(\bar{e}_{x_1} - \bar{e}_{x_2}\right) - \beta_{yz}\,\bar{e}_{z_1}\right]^2.$$

Expanding the square and applying expectation we have:

$$MSE\left(t_{39(2)}\right) \approx \theta_2 \bar{Y}^2 C_y^2 + \left(\theta_2 - \theta_1\right)\beta_{yx}^2 \bar{X}^2 C_x^2 + \theta_1 \beta_{yz}^2 \bar{Z}^2 C_z^2$$
$$+2\left(\theta_1 - \theta_2\right)\beta_{yx}\bar{Y} C_y \bar{X} C_x \rho_{xy} - 2\theta_1 \beta_{yz} \bar{Y} C_y \bar{Z} C_z \rho_{yz} .$$

Substituting the values of regression weights and simplifying, the mean square error of $t_{39(2)}$ is:

$$MSE\left(t_{39(2)}\right) \approx \bar{Y}^2 C_y^2 \left[\theta_2\left(1-\rho_{xy}^2\right) + \theta_1\left(\rho_{xy}^2 - \rho_{yz}^2\right)\right]. \tag{4.7.2}$$

The mean square error given in (4.7.2) is much simpler as compared with mean square error of $t_{37(2)}$ given in (4.6.5).

Samiuddin and Hanif (2006) have also proposed a modification of the Sahoo et al. (1993) estimator by interchanging the role of variables. The modified estimator is:

$$t_{40(2)} = \bar{y}_2 + \beta_{yz}\left(\bar{z}_1 - \bar{z}_2\right) + \beta_{yx}\left(\bar{X} - \bar{x}_1\right). \tag{4.7.3}$$

The mean square error of $t_{40(2)}$ can be easily written from (4.7.2) as:

$$MSE\left(t_{40(2)}\right) \approx \bar{Y}^2 C_y^2 \left[\theta_2\left(1-\rho_{yz}^2\right) + \theta_1\left(\rho_{yz}^2 - \rho_{xy}^2\right)\right]. \tag{4.7.4}$$

Sahoo, Sahoo and Mohanty (1994) have proposed some more regression type estimators in two-phase sampling using information of two auxiliary variables. We discuss these estimators in the following.

4.7.2 Sahoo-Sahoo-Mohanty (1994) Regression-Type Estimator-II

Sahoo et al. (1994) proposed some regression type estimators following the idea of Kiregyera (1984). The first estimator proposed by Sahoo et al. (1994) is same as the estimator $t_{37(2)}$ and we reproduce that estimator here for ready reference as:

$$t_{41(2)} = \bar{y}_2 + \beta_{yx}\left(\bar{x}_1 - \bar{x}_2\right) + \beta_{yz}\left(\bar{Z} - \bar{z}_2\right).$$

The mean square error is given in (4.6.5).

4.7.3 Sahoo-Sahoo-Mohanty (1994) Regression-Type Estimator-III

The third estimator proposed by Sahoo et al. (1994) is slightly different from Kiregyera's (1984) estimator and is given as:

$$t_{42(2)} = \bar{y}_2 + \beta_{yx}\left\{\left(\bar{x}_1 - \bar{x}_2\right) - \beta_{xz}\left(\bar{z}_1 - \bar{z}_2\right) + \beta_{xz}\left(\bar{Z} - \bar{z}_1\right)\right\}. \qquad (4.7.5)$$

The estimator $t_{42(2)}$ may alternately be written as:

$$t_{42(2)} = \bar{y}_2 + \beta_{yx}\left(\bar{x}_1 - \bar{x}_2\right) - \beta_{yx}\beta_{xz}\left(\bar{z}_1 - \bar{z}_2\right) + \beta_{yx}\beta_{xz}\left(\bar{Z} - \bar{z}_1\right).$$

The mean square error of $t_{42(2)}$ may be obtained by written $\bar{y}_2 = \bar{Y} + \bar{e}_{y_2}$ etc.

$$t_{42(2)} - \bar{Y} = \bar{e}_{y_2} + \beta_{yx}\left(\bar{e}_{x_1} - \bar{e}_{x_2}\right) - \beta_{yx}\beta_{xz}\left(\bar{e}_{z_1} - \bar{e}_{z_2}\right) - \beta_{yx}\beta_{xz}\bar{e}_{z_1}.$$

The mean square error is therefore:

$$MSE\left(t_{42(2)}\right) \approx E\left[\bar{e}_{y_2} + \beta_{yx}\left(\bar{e}_{x_1} - \bar{e}_{x_2}\right) - \beta_{yx}\beta_{xz}\left(\bar{e}_{z_1} - \bar{e}_{z_2}\right) - \beta_{yx}\beta_{xz}\bar{e}_{z_1}\right]^2.$$

Expanding the square, applying expectation and using (1.4.1), the mean square error of $t_{42(2)}$ is given as:

$$MSE\left(t_{42(2)}\right) \approx \bar{Y}^2 C_y^2\left[\theta_2 + \theta_1\rho_{xy}^2\rho_{xz}^2 - \left(\theta_2 - \theta_1\right)\left(\rho_{xy}^2\left(1 - \rho_{xz}^2\right) - 2\rho_{xy}\rho_{yz}\rho_{xz}\right)\right]. \qquad (4.7.8)$$

4.7.4 Sahoo-Sahoo-Mohanty (1994) Regression-Type Estimator-IV

The fourth estimator proposed by Sahoo et al. (1994) is:

$$t_{43(2)} = \bar{y}_2 + \beta_{yx}\left\{\left(\bar{x}_1 - \bar{x}_2\right) + \beta_{xz}\left(\bar{Z} - \bar{z}_2\right) + \beta_{xz}\left(\bar{Z} - \bar{z}_1\right)\right\}. \qquad (4.7.9)$$

or

$$t_{43(2)} = \bar{y}_2 + \beta_{yx}\left(\bar{x}_1 - \bar{x}_2\right) - \beta_{yx}\beta_{xz}\left(\bar{z}_2 - \bar{Z}\right) + \beta_{yx}\beta_{xz}\left(\bar{Z} - \bar{z}_1\right).$$

The estimator $t_{43(2)}$ is slight modification of $t_{42(2)}$. The mean square error of $t_{43(2)}$ may immediately be written as:

$$MSE\left(t_{43(2)}\right) \approx \bar{Y}^2 C_y^2 \left[\theta_2 - \left(\theta_2 - \theta_1\right)\left(\rho_{xy}^2 + \rho_{xy}^2\rho_{xz}^2 - 2\rho_{xy}\rho_{yz}\rho_{xz}\right)\right]. \tag{4.7.10}$$

Comparing (4.7.10) with (4.6.3) we can see that $MSE\left(t_{43(2)}\right) < MSE\left(t_{36(2)}\right)$.

Samiuddin and Hanif (2006) have proposed two new estimators which are simpler as compared with estimators proposed by Kiregyera (1984) or by Sahoo et al. (1994). We discuss these estimators in the following.

4.7.5 Samiuddin and Hanif (2006) Estimators

Samiuddin and Hanif (2006) have proposed two estimators of population mean using information of two auxiliary variables. The construction of these two estimators is simpler as compared with estimators proposed by Kiregyera (1984) or by Sahoo et al. (1994). The two estimators proposed by Samiuddin and Hanif (2006) are:

$$t_{44(2)} = \bar{y}_2 + \beta_{yx}\left\{\left(\bar{X} - \bar{x}_2\right) + \beta_{xz}\left(\bar{Z} - \bar{z}_1\right)\right\}, \tag{4.7.11}$$

and

$$t_{45(2)} = \bar{y}_2 + \beta_{yx}\left\{\left(\bar{X} - \bar{x}_1\right) + \beta_{xz}\left(\bar{Z} - \bar{z}_2\right)\right\}. \tag{4.7.12}$$

The mean square errors of the two estimators are:

$$MSE\left(t_{44(2)}\right) = \bar{Y}^2 C_y^2 \left[\theta_2\left(1 - \rho_{xy}^2\right) + 2\theta_1\rho_{xy}^2\rho_{xz}^2 + \theta_1\left(\rho_{xy} - \rho_{yz}\rho_{xz}\right)^2 - \theta_1\rho_{xz}^2\right], \tag{4.7.13}$$

and

$$MSE\left(t_{45(2)}\right) = \bar{Y}^2 C_y^2 \left[\theta_2 - \theta_1\rho_{xy}^2\left(1 - \rho_{xz}^2\right) + \theta_2\left(\rho_{yz}^2\rho_{xz}^2 - 2\rho_{xy}\rho_{yz}\rho_{xz}\right)\right]. \tag{4.7.14}$$

4.7.6 Empirical Study

Samiuddin and Hanif (2006) have used sixteen natural populations for empirical comparison of estimators discussed in sections 4.7.1 through 4.7.3. The empirical study has been conducting by computing mean square errors of estimators $t_{39(2)}$ to $t_{45(2)}$. Samiuddin and Hanif (2006) have used ranked mean square error to decide the best and worst estimators in this group. The ranked mean square errors are given in Tables A.13 through A.15. Table A.13 comprises ranked mean square error of various estimators when all information from the population is used. This table clearly shows that the estimator $t_{44(2)}$ is best in this class and is followed by $t_{39(2)}$. The estimator $t_{42(2)}$ is worst in this class. Tables A.14 and A.15 contain ranked mean square error when sample information is used. These tables show that when information from both phases or only from the second phase is used, then estimator $t_{39(2)}$ is best and is followed by $t_{44(2)}$. The estimator $t_{42(2)}$ is worst in these two situations also. Overall we can conclude that estimators $t_{39(2)}$ and $t_{42(2)}$ are better performers, and $t_{44(2)}$ is worst in the group of estimators.

4.8 Upadhyaya and Singh (2001) Estimators and Some Modifications

We have discussed several estimators for population mean for two-phase sampling which are based upon information of one or two auxiliary variables. The estimators that we have discussed so far have a common property that all estimators are based upon means of variables of interest and auxiliary variables. Different estimators have been proposed in the literature that use information of other summary measures, like coefficient of Skewness, coefficient of kurtosis etc., of auxiliary variables. One such group of estimators has been proposed by Upadhyaya and Singh (2001) which is based upon coefficient of variation and coefficient of kurtosis of auxiliary variables.

Upadhyaya and Singh (2001) proposed four different estimators. The first of these four estimators is:

$$t_{46(2)} = \bar{y}_2\left(\frac{\bar{x}_1}{\bar{x}_2}\right)\left(\frac{\bar{Z}\beta_2(z)+C_z}{\bar{z}_1\beta_2(z)+C_z}\right), \tag{4.8.1}$$

where $\beta_2(z)$ is coefficient of kurtosis and C_z is coefficient of variation of auxiliary variable Z. The mean square error of estimator $t_{46(2)}$ is derived below:

Using notations from section 1.4, we can write:

$$t_{46(2)} - \bar{Y} = \bar{e}_{y_2} + \frac{\bar{Y}}{\bar{X}}\left(\bar{e}_{x_1} - \bar{e}_{x_2}\right) - \left(\frac{\bar{Y}\beta_2(z)}{\bar{Z}\beta_2(z)+C_z}\right)\bar{e}_{z_1}.$$

Squaring and applying expectations, the mean square error is:

$$MSE\left(t_{46(2)}\right) \approx E\left[\bar{e}_{y_2} + \frac{\bar{Y}}{\bar{X}}\left(\bar{e}_{x_1} - \bar{e}_{x_2}\right) - \left(\frac{\bar{Y}\beta_2(z)}{\bar{Z}\beta_2(z)+C_z}\right)\bar{e}_{z_1}\right]^2. \quad (4.8.2)$$

Expanding (4.8.2) and using (1.4.1), the mean square error of $t_{46(2)}$ is:

$$MSE\left(t_{46(2)}\right) \approx \bar{Y}^2\left[\theta_2 C_y^2 + (\theta_2 - \theta_1)\left\{\left(C_x - C_y\rho_{yx}\right)^2 - C_y^2\rho_{yx}^2\right\}\right.$$
$$\left. +\theta_1\left\{\left(\left(1 + C_z/\bar{Z}\beta_2(z)\right)^{-1} C_z - C_y\rho_{zy}\right)^2 - C_y^2\rho_{zy}^2\right\}\right]. \quad (4.8.3)$$

Samiuddin and Hanif (2006) have proposed a modification of $t_{46(2)}$ by interchanging the variables. The modified estimator is:

$$t_{47(2)} = \bar{y}_2\left(\frac{\bar{z}_1}{\bar{z}_2}\right)\left(\frac{\bar{X}\beta_2(x)+C_x}{\bar{x}_1\beta_2(x)+C_x}\right), \quad (4.8.4)$$

The mean square error of estimator (4.8.4) may be easily written following (4.8.3). Upadhyaya and Singh (2001) have proposed three more estimators along the lines of $t_{46(2)}$. Samiuddin and Hanif (2006) have provided modification of each of three estimators of Upadhyaya and Singh (2001), i.e.:

$$t_{48(2)} = \bar{y}_2\left(\frac{\bar{x}_1}{\bar{x}_2}\right)\left\{\frac{\bar{Z}C_z + \beta_2(z)}{\bar{z}_1 C_z + \beta_2(z)}\right\}, \quad (4.8.5)$$

(Upadhyaya and Singh, 2001)

$$t_{49(2)} = \bar{y}_2\left(\frac{\bar{z}_1}{\bar{z}_2}\right)\left\{\frac{\bar{X}C_x + \beta_2(x)}{\bar{x}_1 C_x + \beta_2(x)}\right\}, \quad (4.8.6)$$

(Samiuddin and Hanif, 2006)

$$t_{50(2)} = \bar{y}_2\left(\frac{\bar{x}_1}{\bar{x}_2}\right)\left\{\frac{\bar{z}_1\beta_2(z)+C_z}{\bar{Z}\beta_2(z)+C_z}\right\}, \quad (4.8.7)$$

(Upadhyaya and Singh, 2001)

$$t_{51(2)} = \bar{y}_2 \left(\frac{\bar{z}_1}{\bar{z}_2}\right)\left\{\frac{\bar{x}_1 \beta_2(x) + C_x}{\bar{X}\beta_2(x) + C_x}\right\}, \tag{4.8.8}$$

(Samiuddin and Hanif, 2006)

$$t_{52(2)} = \bar{y}_2 \left(\frac{\bar{x}_1}{\bar{x}_2}\right)\left\{\frac{\bar{z}_1 C_z + \beta_2(z)}{\bar{Z} C_z + \beta_2(z)}\right\}, \tag{4.8.9}$$

(Upadhyaya and Singh, 2001)

and

$$t_{53(2)} = \bar{y}_2 \left(\frac{\bar{z}_1}{\bar{z}_2}\right)\left\{\frac{\bar{x}_1 C_x + \beta_2(x)}{\bar{X} C_x + \beta_2(x)}\right\}. \tag{4.8.10}$$

(Samiuddin and Hanif, 2006)

Upadhyaya and Singh (2001) have shown that the mean square errors of their proposed estimators may be derived along the lines of the mean square error of $t_{46(2)}$. We can easily show that the mean square errors of $t_{48(2)}$, $t_{50(2)}$ and $t_{52(2)}$ are:

$$MSE\left(t_{48(2)}\right) \approx \bar{Y}^2 \left[\theta_2 C_y^2 + (\theta_2 - \theta_1)\left\{\left(C_x - C_y \rho_{xy}\right)^2 - C_y^2 \rho_{xy}^2\right\} + \theta_1 \left\{\left(\left(1 + \beta_2(z)/\bar{Z} C_z\right)^{-1} C_z - C_y \rho_{yz}\right)^2 - C_y^2 \rho_{yz}^2\right\}\right], \tag{4.8.11}$$

$$MSE\left(t_{50(2)}\right) \approx \bar{Y}^2 \left[\theta_2 C_y^2 + (\theta_2 - \theta_1)\left\{\left(C_x - C_y \rho_{xy}\right)^2 - C_y^2 \rho_{xy}^2\right\} + \theta_1 \left\{\left(\left(1 + C_z/\bar{Z}\beta_2(z)\right)^{-1} C_z + C_y \rho_{yz}\right)^2 - C_y^2 \rho_{yz}^2\right\}\right]. \tag{4.8.12}$$

$$MSE\left(t_{52(2)}\right) \approx \bar{Y}^2 \left[\theta_2 C_y^2 + (\theta_2 - \theta_1)\left\{\left(C_x - C_y \rho_{xy}\right)^2 - C_y^2 \rho_{xy}^2\right\} + \theta_1 \left\{\left(\left(1 + \beta_2(z)/\bar{Z} C_z\right)^{-1} C_z + C_y \rho_{yz}\right)^2 - C_y^2 \rho_{yz}^2\right\}\right]. \tag{4.8.13}$$

The mean square errors of the estimators suggested by Samiuddin and Hanif (2006) may be directly written from (4.8.11), (4.8.12) and (4.8.13) by changing subscript x with z.

4.8.1 Empirical Study

Samiuddin and Hanif (2006) have empirically compared Upadhyaya and Singh (2001) estimators and associated modifications. The empirical comparison is based upon sixteen natural populations. Samiuddin and Hanif (2006) have conducted the empirical comparison by using ranked mean square errors of estimators $t_{46(2)}$ to $t_{53(2)}$. We have given the ranked mean square errors in Tables A.16 to A.18. Looking at table A.16 we can see that the estimator $t_{46(2)}$ is best in this group when all measures involved in the mean square error are computed from the population. The estimator $t_{48(2)}$ is second best in this situation, while the estimator $t_{53(2)}$ is worst. Table A.17 contains ranked mean square errors of Upadhyaya and Singh (2001) estimators and associated modifications when measures are computed using information from both phases. This table shows that the estimator $t_{46(2)}$ is again best and is again followed by $t_{48(2)}$. The estimator $t_{50(2)}$ is worst in this situation. When all measures are computed from second phase sample then, from Table A.18, it can be seen that the performance of estimator $t_{48(2)}$ is best and is followed by $t_{49(2)}$. The estimator $t_{52(2)}$ has worst performance in this case.

4.9 Roy's (2003) Regression Estimator

We have discussed some regression type estimators in two phase sampling which are proposed by Kiregyera (1984), Mukerjee et al. (1987) and Sahoo et al. (1993) in sections 4.5, 4.6 and 4.7 respectively. Roy (2003), following the idea of these proposed estimators, suggested his regression type estimator based upon information of two auxiliary variables. The proposed estimator is:

$$t_{54(2)} = \bar{y}_2 + k_1\left[\left\{\bar{z}_1 + k_2\left(\bar{X} - \bar{x}_1\right)\right\} - \left\{\bar{z}_2 + k_3\left(\bar{X} - \bar{x}_2\right)\right\}\right], \qquad (4.9.1)$$

where k_1, k_2 and k_3 are constants to be determined so that mean square error of (4.9.1) is minimum. Using section 1.4, we can write (4.9.1) as:

$$t_{54(2)} - \bar{Y} = \bar{e}_{y_2} + k_1\left(\bar{e}_{z_1} - \bar{e}_{z_2}\right) - k_1k_2\bar{e}_{x_1} + k_1k_3\bar{e}_{x_2},$$

$$\text{or} \quad t_{54(2)} - \bar{Y} = \bar{e}_{y_2} + \alpha\left(\bar{e}_{z_1} - \bar{e}_{z_2}\right) - \beta\bar{e}_{x_1} - \gamma\bar{e}_{x_2}, \qquad (4.9.2)$$

where $\alpha = k_1$, $\beta = k_1 k_2$ and $\gamma = -k_1 k_3$. Squaring (4.9.2) and applying expectation, the mean square error of $t_{54(2)}$ is:

$$MSE\left(t_{54(2)}\right) = E\left[\bar{e}_{y_2} + \alpha\left(\bar{e}_{z_1} - \bar{e}_{z_2}\right) - \beta\bar{e}_{x_1} - \gamma\bar{e}_{x_2}\right]^2. \tag{4.9.3}$$

In order to obtain optimum values of an unknown, we differentiate (4.9.3) *w.r.t.* each and equate the derivative to zero. The estimating equations, thus obtained, are:

$$\alpha \bar{Z} C_z + \gamma \bar{X} C_x \rho_{xz} = YC_y \rho_{yz}, \tag{4.9.4}$$

$$\beta \bar{X} C_z + \gamma \bar{X} C_x = \bar{Y} C_y \rho_{xy}, \tag{4.9.5}$$

and

$$\alpha \bar{Z} C_z \rho_{xz} - \beta \bar{X} C_x = 0. \tag{4.9.6}$$

Solving (4.9.4) to (4.9.6) simultaneously we have:

$$\alpha = k_1 = \frac{\bar{Y}\, C_y}{\bar{Z}\, C_z} \cdot \frac{\rho_{yz} - \rho_{xz}\rho_{xy}}{1-\rho_{xz}^2} = \beta_{yz.x}, \tag{4.9.7}$$

$$\beta = k_1 k_2 = \frac{\bar{Y}\, C_y}{\bar{X}\, C_x} \cdot \frac{\rho_{xz}\left(\rho_{yz} - \rho_{xy}\rho_{xz}\right)}{1-\rho_{xz}^2}, \tag{4.9.8}$$

and

$$\gamma = -k_1 k_3 = \frac{\bar{Y}\, C_y}{\bar{X}\, C_x} \cdot \frac{\rho_{xy} - \rho_{yz}\rho_{xz}}{1-\rho_{xz}^2}. \tag{4.9.9}$$

Now, from (4.9.7) and (4.9.8) we have:

$$k_2 = \beta/k_1 = \frac{\bar{Z} C_z}{\bar{X} C_x} \rho_{xz} = \beta_{zx}. \tag{4.9.10}$$

Again from (4.9.7) and (4.9.9) we have:

$$k_3 = -\gamma/k_1 = -\frac{\bar{Z} C_z\left(\rho_{xy} - \rho_{yz}\rho_{xz}\right)}{\bar{X}\, C_x\left(\rho_{yz} - \rho_{xz}\rho_{yz}\right)} = \beta_{zx} - \frac{\beta_{yx}}{\beta_{yz.x}}. \tag{4.9.11}$$

Now using (4.9.7), (4.9.10) and (4.9.11) in (4.9.3) and simplifying, the mean square error of $t_{54(2)}$ is:

$$MSE\left(t_{54(2)}\right) \approx \bar{Y}^2 C_y^2\left[\theta_2\left(1-\rho_{y.xz}^2\right) + \theta_1 \rho_{yz.x}^2\left(1-\rho_{xy}^2\right)\right], \tag{4.9.12}$$

where $\rho_{y.xz}^2$ is squared multiple correlation between Y and combined effects of X and Z and $\rho_{yz.x}^2$ is partial correlation between Y and Z after removing effects of X. Roy (2003) has compared $t_{54(2)}$ with several well-known two phase sampling estimators and has shown that his proposed estimator has a smaller mean square error as compared with popular estimators proposed by Kiregyera (1984), Mukharjee et al. (1987) and Sahoo et al. (1993). A modification of $t_{54(2)}$, proposed by Samiuddin and Hanif (2006) is:

$$t_{55(2)} = \bar{y}_2 + k_1\left[\left\{\bar{x}_1 + k_2\left(\bar{Z} - \bar{z}_1\right)\right\} - \left\{\bar{x}_2 + k_3\left(\bar{Z} - \bar{z}_2\right)\right\}\right]. \tag{4.9.13}$$

The mean square error of (4.9.13) is easily written from (4.9.12).

4.10 Hanif-Hamad-Shabaz (2010) Estimators

Hanif et al. (2010) have proposed estimators under various situations depending upon availability of population information of auxiliary variables. The authors have considered all three cases, i.e. no information, partial information, and full information. The proposed estimators by Hanif et al. (2010) under various situations are discussed in the following.

4.10.1 Estimators for No Information Case

Hanif et al. (2010) have proposed the following estimator of population mean when only sample information is available:

$$t_{56(2)} = \left\{\bar{y}_2 + \beta_{yx}\left(\bar{x}_1 - \bar{x}_2\right)\right\}\left\{k\left(\bar{z}_1/\bar{z}_2\right) + \left(1-k\right)\left(\bar{z}_2/\bar{z}_1\right)\right\}, \tag{4.10.1}$$

where k is a constant to be determined so that the mean square error of $t_{56(2)}$ is minimum. Substituting $\bar{y}_2 = \bar{Y} + \bar{e}_{y_2}$ etc. in (4.10.1), and retaining linear terms only, we have:

$$t_{56(2)} - \bar{Y} = \bar{e}_{y_2} + \beta_{yx}\left(\bar{e}_{x_1} - \bar{e}_{x_2}\right) - \left(\bar{Y}/\bar{Z}\right)\left(1-2k\right)\left(\bar{e}_{z_1} - \bar{e}_{z_2}\right).$$

The mean square error of $t_{56(2)}$ is:

$$MSE\left(t_{56(2)}\right) \approx E\left[\bar{e}_{y_2} + \beta_{yx}\left(\bar{e}_{x_1} - \bar{e}_{x_2}\right) - \left(\bar{Y}/\bar{Z}\right)\left(1-2k\right)\left(\bar{e}_{z_1} - \bar{e}_{z_2}\right)\right]^2.$$

Expanding the square and applying expectation, we have:

$$MSE\left(t_{56(2)}\right)\approx\left[\theta_2\bar{Y}^2C_y^2-(\theta_2-\theta_1)\bar{Y}^2C_y^2\rho_{xy}^2+(\theta_2-\theta_1)\bar{Y}^2C_z^2\right.$$
$$+4k^2(\theta_2-\theta_1)\bar{Y}^2C_z^2+2(\theta_2-\theta_1)\bar{Y}^2C_yC_z\rho_{yz}$$
$$-4k(\theta_2-\theta_1)\bar{Y}^2C_yC_z\rho_{yz}-2(\theta_2-\theta_1)\bar{Y}^2C_yC_z\rho_{xy}\rho_{xz}$$
$$\left.+4k(\theta_2-\theta_1)\bar{Y}^2C_yC_z\rho_{xy}\rho_{xz}-4k(\theta_2-\theta_1)\bar{Y}^2C_z^2\right]. \quad (4.10.2)$$

The optimum value of k which minimizes (4.10.2) is:

$$k=\frac{1}{2}\left(1+\frac{C_y}{C_z}\rho_{yz}-\frac{C_y}{C_z}\rho_{xy}\rho_{xz}\right). \quad (4.10.3)$$

Using the value of k in (4.10.2) and simplifying, the mean square error of $t_{56(2)}$ is:

$$MSE\left(t_{56(2)}\right)\approx\bar{Y}^2C_y^2\left[\theta_2-(\theta_2-\theta_1)\left\{\rho_{xy}^2+\left(\rho_{yz}-\rho_{xy}\rho_{xz}\right)^2\right\}\right]. \quad (4.10.4)$$

Hanif et al. (2010) have also proposed a modification of $t_{56(2)}$ as:

$$t_{57(2)}=\left\{\bar{y}_2+k_1\left(\bar{x}_1-\bar{x}_2\right)\right\}\left\{k_2\left(\bar{z}_1/\bar{z}_2\right)+\left(1-k_2\right)\left(\bar{z}_2/\bar{z}_1\right)\right\}, \quad (4.10.5)$$

where k_1 and k_2 are constants to be determined. Using notations from (1.4), we have:

$$t_{57(2)}-\bar{Y}=\bar{e}_{y_2}+k_1\left(\bar{e}_{x_1}-\bar{e}_{x_2}\right)-\left(\bar{Y}/\bar{Z}\right)\left(1-2k_2\right)\left(\bar{e}_{z_1}-\bar{e}_{z_2}\right). \quad (4.10.6)$$

Squaring (4.6.10) and applying expectation, the mean square error of $t_{57(2)}$ is:

$$MSE\left(t_{57(2)}\right)=\left[\theta_2\bar{Y}^2C_y^2+k_1^2\left(\theta_2-\theta_1\right)\bar{X}^2C_x^2+\left(\theta_2-\theta_1\right)\bar{Y}^2C_z^2\right.$$
$$+4k_2^2\left(\theta_2-\theta_1\right)\bar{Y}^2C_z^2-2k_1\left(\theta_2-\theta_1\right)\bar{X}\bar{Y}C_xC_y\rho_{xy}$$
$$+2\left(\theta_2-\theta_1\right)\bar{Y}^2C_yC_z\rho_{yz}-4k_2\left(\theta_2-\theta_1\right)\bar{Y}^2C_yC_z\rho_{yz}$$
$$-2k_1\left(\theta_2-\theta_1\right)\bar{X}\bar{Y}C_xC_z\rho_{xz}$$
$$\left.+4k_1k_2\left(\theta_2-\theta_1\right)\bar{X}\bar{Y}C_xC_z\rho_{xz}-4k_2\left(\theta_2-\theta_1\right)\bar{Y}^2C_z^2\right]. \tag{4.10.7}$$

The optimum values of k_1 and k_2 which minimize (4.7.10) are:

$$k_1=\frac{\bar{Y}C_y}{\bar{X}C_x}\left(\frac{\rho_{xy}-\rho_{xz}\rho_{yz}}{1-\rho_{xz}^2}\right)=\beta_{yx.z}, \tag{4.10.8}$$

and

$$k_2=\frac{1}{2}\left\{1+\frac{C_y}{C_z}\left(\frac{\rho_{yz}-\rho_{xy}\rho_{xz}}{1-\rho_{xz}^2}\right)\right\}=\frac{1}{2}\left(1+\frac{\bar{Y}}{\bar{Z}}\beta_{yz.x}\right). \tag{4.10.9}$$

Using (4.10.8) and (4.10.9) in (4.10.7) and simplifying, the mean square error of $t_{57(2)}$ is:

$$MSE\left(t_{57(2)}\right)=\bar{Y}^2C_y^2\left[\theta_2-\left(\theta_2-\theta_1\right)\left\{1-\left(1-\rho_{xy.z}^2\right)\left(1-\rho_{yz}^2\right)\right\}\right]. \tag{4.10.10}$$

Further, by using $\left(1-\rho_{xy.z}^2\right)\left(1-\rho_{yz}^2\right)=1-\rho_{y.xz}^2$, we can write:

$$MSE\left(t_{57(2)}\right)=\bar{Y}^2C_y^2\left[\theta_2\left(1-\rho_{y.xz}^2\right)+\theta_1\rho_{y.xz}^2\right]. \tag{4.10.11}$$

Comparing (4.10.4) with (4.10.11) we can see that $t_{57(2)}$ is a more efficient estimator than $t_{56(2)}$.

4.10.2 Estimators for Partial Information Case

In this regard Hanif et al. (2010) proposed two estimators when population information of one of the auxiliary variable is known. The first of two proposed estimators is:

$$t_{58(2)}=\left\{\bar{y}_2+\beta_{yx}\left(\bar{X}-\bar{x}_2\right)\right\}\left\{k\left(\bar{z}_1/\bar{z}_2\right)+\left(1-k\right)\left(\bar{z}_2/\bar{z}_1\right)\right\}. \tag{4.10.12}$$

The estimator $t_{58(2)}$ is like $t_{56(2)}$ with the difference that $\bar{x}_1$ is replaced with $\bar{X}$. The mean square error of $t_{58(2)}$ can be readily written as:

$$MSE\left(t_{58(2)}\right) \approx \Big[\theta_2 \bar{Y}^2 C_y^2 - \theta_1 \bar{Y}^2 C_y^2 \rho_{xy}^2 + \left(\theta_2 - \theta_1\right)\bar{Y}^2 C_z^2$$
$$+4k^2\left(\theta_2 - \theta_1\right)\bar{Y}^2 C_z^2 + 2\left(\theta_2 - \theta_1\right)\bar{Y}^2 C_y C_z \rho_{yz}$$
$$-4k\left(\theta_2 - \theta_1\right)\bar{Y}^2 C_y C_z \rho_{yz} - 2\left(\theta_2 - \theta_1\right)\bar{Y}^2 C_y C_z \rho_{xy}\rho_{xz}$$
$$+4k\left(\theta_2 - \theta_1\right)\bar{Y}^2 C_y C_z \rho_{xy}\rho_{xz} - 4k\left(\theta_2 - \theta_1\right)\bar{Y}^2 C_z^2\Big]. \quad (4.10.13)$$

The optimum value of k which minimizes (4.10.13) is:

$$k = \frac{1}{2}\left(1 + \frac{C_y}{C_z}\rho_{yz} - \frac{C_y}{C_z}\rho_{xy}\rho_{xz}\right) = \frac{1}{2}\left\{1 + \frac{\bar{Y}}{\bar{Z}}\beta_{yz.x}\left(1 - \rho_{xz}^2\right)\right\}. \quad (4.10.14)$$

Using (4.10.14) in (4.10.13) and simplifying, the mean square error of $t_{58(2)}$ is:

$$MSE\left(t_{58(2)}\right) \approx \bar{Y}^2 C_y^2\left[\theta_2\left(1 - \rho_{xy}^2\right) - \left(\theta_2 - \theta_1\right)\left(\rho_{yz} - \rho_{xy}\rho_{xz}\right)^2\right]. \quad (4.10.15)$$

The second estimator for partial information case, proposed by Hanif et al. (2010) is:

$$t_{59(2)} = \left\{\bar{y}_2 + \beta_{yx}\left(\bar{x}_1 - \bar{x}_2\right)\right\}\left\{k\left(\bar{Z}/\bar{z}_1\right) + \left(1-k\right)\left(\bar{z}_1/\bar{Z}\right)\right\}. \quad (4.10.16)$$

The optimum value of k which minimizes the mean square error of $t_{59(2)}$ is:

$$k = \frac{1}{2}\left(1 + \frac{C_y}{C_z}\rho_{yz}\right) = \frac{1}{2}\left(1 + \frac{\bar{Y}}{\bar{Z}}\beta_{yz}\right), \quad (4.10.17)$$

and minimum mean square error of $t_{59(2)}$ is:

$$MSE\left(t_{59(2)}\right) \approx \bar{Y}^2 C_y^2\left[\theta_2\left(1 - \rho_{xy}^2\right) - \theta_1\left(\rho_{yz}^2 - \rho_{xy}^2\right)\right]. \quad (4.10.18)$$

The mean square error of $t_{59(2)}$ given in (4.10.18) is exactly same as the mean square error of the Sahoo et al. (1993) estimator given in (4.7.2).

4.10.3 Estimators for Full Information Case

Hanif et al. (2010) have also proposed two estimators when population information of both auxiliary variables is available. The first of two estimators is:

$$t_{60(2)} = \left\{\bar{y} + \beta_{yx}\left(\bar{X} - \bar{x}_2\right)\right\}\left\{k\left(\bar{Z}/\bar{z}_2\right) + (1-k)\left(\bar{z}_2/\bar{Z}\right)\right\}. \tag{4.10.19}$$

The estimator $t_{60(2)}$ is like $t_{56(2)}$ with the difference that $\bar{x}_1$ has been replaced with $\bar{X}$ and $\bar{z}_1$ has been replaced with $\bar{Z}$. It has been shown that the optimum value of k which minimizes mean square error of $t_{60(2)}$ is same as given in (4.10.14) and the mean square error of $t_{60(2)}$ is:

$$MSE\left(t_{60(2)}\right) \approx \theta_2 \bar{Y}^2 C_y^2\left[\left(1-\rho_{xy}^2\right) - \left(\rho_{yz} - \rho_{xy}\rho_{xz}\right)^2\right]. \tag{4.10.20}$$

The second estimator proposed by Hanif et al. (2010) for the full information case is a slight modification of $t_{60(2)}$ and is:

$$t_{61(2)} = \left\{\bar{y} + k_1\left(\bar{X} - \bar{x}_2\right)\right\}\left\{k_2\left(\bar{Z}/\bar{z}_2\right) + (1-k_2)\left(\bar{z}_2/\bar{Z}\right)\right\}, \tag{4.10.21}$$

where k_1 and k_2 are constants to be determined. Using $\bar{y}_2 = \bar{Y} + \bar{e}_{y_2}$ etc. we can write:

$$t_{61(2)} - \bar{Y} = \bar{e}_{y_2} - k_1\bar{e}_{x_2} - \frac{\bar{Y}}{\bar{Z}}\bar{e}_{z_2}\left(1-2k_2\right),$$

and hence the mean square error of $t_{61(2)}$ is:

$$MSE\left(t_{61(2)}\right) = E\left[\bar{e}_{y_2} - k_1\bar{e}_{x_2} - \frac{\bar{Y}}{\bar{Z}}\bar{e}_{z_2}\left(1-2k_2\right)\right]^2. \tag{4.10.22}$$

The optimum values of k_1 and k_2 which minimizes (4.10.22) are the same as given in (4.10.8) and (4.10.9). Using optimum values of k_1 and k_2 in (4.10.22) and simplifying, the mean square error of $t_{61(2)}$ is:

$$MSE\left(t_{61(2)}\right) = \theta_2 \bar{Y}^2 C_y^2\left(1-\rho_{yx.z}^2\right)\left(1-\rho_{yz}^2\right), \tag{4.10.23}$$

which can alternatively be written, by using $\left(1-\rho_{xy.z}^{2}\right)\left(1-\rho_{yz}^{2}\right)=1-\rho_{y.xz}^{2}$, as:

$$MSE\left(t_{61(2)}\right)=\theta_{2}\bar{Y}^{2}C_{y}^{2}\left(1-\rho_{y.xz}^{2}\right). \tag{4.10.24}$$

CHAPTER FIVE

GENERAL FAMILIES OF ESTIMATORS IN TWO-PHASE SAMPLING

5.1 Introduction

We have seen that the two-phase sampling is very useful in estimation of population mean or total when the sampling frame is not known. The sample collected at the first-phase serves as the basis of additional information which can be effectively incorporated in ratio and regression methods of estimation. We have discussed several two-phase sampling estimators which are based upon one and two auxiliary variables. In this Chapter we will discuss various general estimators for two-phase sampling which are based upon one and two auxiliary variables. The estimators discussed in this Chapter are an extension to estimators discussed in the previous chapter. We start by presenting an extension of the classical ratio estimator of two phase sampling given by Srivastava (1971).

5.2 Srivastava's (1971) Generalized Estimator using One Auxiliary Variable

Srivastava (1971) proposed a general estimator for two-phase sampling using information of a single auxiliary variable. The proposed estimator is:

$$t_{62(2)} = \bar{y}_2 \left(\frac{\bar{x}_1}{\bar{x}_2} \right)^{\alpha}, \tag{5.2.1}$$

where α is a constant. The classical ratio estimator for two-phase sampling immediately emerges from $t_{62(2)}$ for $\alpha = 1$. Using section 1.4 the estimator $t_{62(2)}$ may be written as:

$$t_{62(2)} - \bar{Y} = \bar{e}_{y_2} + \frac{\bar{Y}}{\bar{X}} \alpha \left(\bar{e}_{x_1} - \bar{e}_{x_2} \right).$$

The mean square error is:

$$MSE\left(t_{62(2)}\right)=E\left[\bar{e}_{y_2}+\frac{\bar{Y}}{\bar{X}}\alpha\left(\bar{e}_{x_1}-\bar{e}_{x_2}\right)\right]^2. \tag{5.2.2}$$

Differentiating (5.2.2) *w.r.t.* α and equating to zero, the optimum value of α is $\alpha=\left(C_y/C_x\right)\rho_{xy}$. Using optimum value of α in (5.2.2) and simplifying, the mean square error of $t_{62(2)}$ is:

$$MSE\left(t_{62(2)}\right) \approx \bar{Y}^2C_y^2\left[\theta_2\left(1-\rho_{xy}^2\right)+\theta_1\rho_{xy}^2\right]. \tag{5.2.3}$$

The mean square error given in (5.2.3) is same as the mean square error of the classical regression estimator of two-phase sampling given in (4.2.11). We, therefore, say that the Srivastava (1971) general estimator is in equivalence class with the classical regression estimator of two-phase sampling. Bedi (1985) proposed independently his general estimator which is the same as the Srivastava (1971) estimator but with a different auxiliary variable. The estimator proposed by Bedi (1985) is:

$$t_{63(2)}=\bar{y}_2\left(\frac{\bar{z}_1}{\bar{z}_2}\right)^{\alpha}.$$

The mean square error of $t_{63(2)}$ is the same as given in (5.2.3) with the subscripts interchanged.

The general estimators in two-phase sampling based upon information of two auxiliary variables will be discussed.

5.3 Generalized Estimators using Two Auxiliary Variables

In previous chapters we have discussed several estimators for two-phase sampling that use the information of two auxiliary variables. Now we will discuss various general estimators for two-phase sampling which are based upon two auxiliary variables.

5.3.1 Srivastava-Khare -Srivastava (1990) Chain-Ratio-Type Estimator

Srivastava et al. (1990) extended the chain ratio estimator proposed by Chand (1975). The estimator for two auxiliary variables is:

$$t_{64(2)} = \bar{y}_2 \left(\frac{\bar{x}_1}{\bar{x}_2} \right)^{\alpha_1} \left(\frac{\bar{Z}}{\bar{z}_1} \right)^{\alpha_2}, \quad (5.3.1)$$

where α_1 and α_2 are constants to be determined. The estimator $t_{62(2)}$ reduces to chain ratio estimator proposed by Chand (1975) for $\alpha_1 = \alpha_2 = 1$. Substituting $\bar{y}_2 = \bar{Y} + \bar{e}_{y_2}$ etc. and expanding up to linear term only, we have:

$$t_{64(2)} - \bar{Y} = \bar{e}_{y_2} + \alpha_1 \frac{\bar{Y}}{\bar{X}} \left(\bar{e}_{x_1} - \bar{e}_{x_2} \right) - \alpha_2 \frac{\bar{Y}}{\bar{Z}} \bar{e}_{z_1}.$$

Squaring and applying expectation, the mean square error of $t_{64(2)}$ is:

$$MSE\left(t_{64(2)}\right) = E\left[\bar{e}_{y_2} + \alpha_1 \frac{\bar{Y}}{\bar{X}} \left(\bar{e}_{x_1} - \bar{e}_{x_2} \right) - \alpha_2 \frac{\bar{Y}}{\bar{Z}} \bar{e}_{z_1} \right]^2. \quad (5.3.2)$$

The optimum values of α_1 and α_2 which minimizes (5.3.2) are:

$$\alpha_1 = \left(C_y / C_x \right) \rho_{xy} \text{ and } \alpha_2 = \left(C_y / C_z \right) \rho_{yz}.$$

Using optimum values of $g_\omega\left(\bar{y}_2, v_1, v_{2d}\right) = \frac{\bar{y}_2}{k_1 + k_2} \left[k_1 v_1^{\frac{\lambda_1}{2}} + k_2 e^{\lambda_2 \left(v_{2d} - 1 \right)} \right]$

and $g_\omega\left(\bar{y}_2, v_1, v_{2d}\right) = \bar{y}_2 \left[k e^{\lambda_1 \left(v_1 - 1 \right)} + \left(1 - k \right) e^{\lambda_2 \left(v_{2d} - 1 \right)} \right]$ in (5.3.2) and simplifying, the mean square error of $t_{64(2)}$ is:

$$MSE\left(t_{64(2)}\right) \approx \bar{Y}^2 C_y^2 \left[\theta_2 \left(1 - \rho_{xy}^2 \right) + \theta_1 \left(\rho_{xy}^2 - \rho_{yz}^2 \right) \right]. \quad (5.3.3)$$

The mean square error given in (5.3.3) is the same as the mean square error of the Sahoo et al. (1993) estimator given in (4.7.1). A modification of $t_{64(2)}$, proposed by Samiuddin and Hanif (2006), is:

$$t_{65(2)} = \bar{y}_2 \left(\frac{\bar{z}_1}{\bar{z}_2} \right)^{\alpha_1} \left(\frac{\bar{X}}{\bar{x}_1} \right)^{\alpha_2}. \quad (5.3.4)$$

The mean square error of (5.3.4) may immediately be written from (5.3.3) with the change of subscripts.

5.3.2 Sahoo and Sahoo (1994) Regression-cum-Ratio Generalized Estimator

Sahoo and Sahoo (1994) proposed the following general regression-cum-ratio estimator:

$$t_{66(2)} = \left\{\bar{y}_2 + \beta_{yx}\left(\bar{X} - \bar{x}_1\right)\right\}\left(\frac{\bar{z}_2}{\bar{Z}}\right)^{\alpha}. \qquad (5.3.5)$$

The estimator $t_{13(2)}$ proposed by Ahmad et al. (2007) emerges as a special case of $t_{66(2)}$ for $\alpha = 1$. Using $\bar{y}_2 = \bar{Y} + \bar{e}_{y_2}$ etc. in (5.3.5) we have:

$$t_{66(2)} - \bar{Y} = \bar{e}_{y_2} - \beta_{yx}\bar{e}_{x_1} + \alpha\left(\bar{Y}/\bar{Z}\right)\bar{e}_{z_2}.$$

Squaring and applying expectation, we have:

$$MSE\left(t_{66(2)}\right) = E\left[\bar{e}_{y_2} - \beta_{yx}\bar{e}_{x_1} + \alpha\left(\bar{Y}/\bar{Z}\right)\bar{e}_{z_2}\right]^2. \qquad (5.3.6)$$

The optimum value of α which minimizes (5.3.6) is:

$$\alpha = \frac{C_y}{C_z}\rho_{yz} + \frac{\theta_1}{\theta_2}\frac{C_y}{C_z}\rho_{xy}\rho_{xz}. \qquad (5.3.7)$$

Using (5.3.7) in (5.3.6) and simplifying, the mean square error of $t_{66(2)}$ becomes:

$$MSE\left(t_{66(2)}\right) \approx \bar{Y}^2 C_y^2\left[\theta_2\left(1-\rho_{yz}^2\right) - \left(\theta_1/\theta_2\right)\rho_{xy}^2\left(\theta_2 + \theta_1\rho_{xz}^2\right) + 2\theta_1\rho_{xy}\rho_{yz}\rho_{xz}\right]. \qquad (5.3.8)$$

The mean square error given in (5.3.8) is larger than mean square error of $t_{64(2)}$ given in (5.3.3).

5.3.3 Ahmad et al. (2007) Generalized Chain Ratio Estimators

Ahmad et al. (2007) have proposed two estimators on the lines of Chand (1975). The first of two proposed estimators is:

$$t_{67(2)} = \bar{y}_2\left(\frac{\bar{x}_1}{\bar{x}_2}\right)^{\alpha_1}\left(\frac{\bar{Z}}{\bar{z}_1}\right)^{\alpha_2}\left(\frac{\bar{Z}}{\bar{z}_2}\right)^{\alpha_3}, \qquad (5.3.9)$$

where α_1, α_2 and α_3 are constants. Following notations from section 1.4, the estimator $t_{67(2)}$ is written as:

$$t_{67(2)} - \bar{Y} = \bar{e}_{y_2} + \alpha_1 \frac{\bar{Y}}{\bar{\bar{X}}}\left(\bar{e}_{x_1} - \bar{e}_{x_2}\right) - \alpha_2 \frac{\bar{Y}}{\bar{\bar{Z}}}\bar{e}_{z_1} - \alpha_3 \frac{\bar{Y}}{\bar{\bar{Z}}} Y\bar{e}_{z_2}.$$

Squaring and applying expectation, the mean square error of $t_{67(2)}$ is:

$$MSE\left(t_{67(2)}\right) = E\left[\bar{e}_{y_2} + \frac{\bar{Y}}{\bar{\bar{X}}}\alpha_1\left(\bar{e}_{x_1} - \bar{e}_{x_2}\right) - \alpha_2 \frac{\bar{Y}}{\bar{\bar{Z}}}\bar{e}_{z_1} - \alpha_3 \frac{\bar{Y}}{\bar{\bar{Z}}}\bar{e}_{z_2}\right]^2. \quad (5.3.10)$$

The normal equations for obtaining optimum values of α_1, α_2 and α_3 are:

$$\alpha_1 C_x + \alpha_3 C_z \rho_{xz} - C_y \rho_{xy} = 0, \quad (5.3.11)$$

$$\alpha_2 C_z + \alpha_3 C_z - C_y \rho_{yz} = 0, \quad (5.3.12)$$

and

$$\alpha_1\left(\theta_2 - \theta_1\right) C_x \rho_{xz} + \alpha_2 \theta_1 C_z + \alpha_3 \theta_2 C_z - \theta_2 C_y \rho_{yz} = 0. \quad (5.3.13)$$

Solving (5.3.11), (5.3.12) and (5.3.13) simultaneously, the optimum values of α_1, α_2 and α_3 are:

$$\alpha_1 = \frac{C_y}{C_x}\left(\frac{\rho_{xy} - \rho_{xy}\rho_{xz}}{1 - \rho^2_{xz}}\right) = \frac{\bar{Y}}{\bar{\bar{X}}}\beta_{yx.z}, \quad (5.3.14)$$

$$\alpha_2 = \frac{C_y}{C_z}\left(\frac{\rho_{yz} - \rho_{yz}\rho^2_{xz} - \rho_{xy} + \rho_{yz}\rho_{xz}}{1 - \rho^2_{xz}}\right) = \frac{\bar{Y}}{\bar{\bar{Z}}}\left(\beta_{yz} - \beta_{yx.z}\right), \quad (5.3.15)$$

and

$$\alpha_3 = \frac{C_y}{C_z}\left(\frac{\rho_{yz} - \rho_{xy}\rho_{xz}}{1 - \rho^2_{xz}}\right) = \frac{\bar{Y}}{\bar{\bar{Z}}}\beta_{yz.x}. \quad (5.3.16)$$

Using (5.3.14), (5.3.15) and (5.3.16) in (5.3.10) and simplifying, the mean square error of $t_{67(2)}$ is:

$$MSE\left(t_{67(2)}\right) \approx \bar{Y}^2 C_y^2\left[\theta_2\left\{1 - \frac{\left(\rho^2_{yx} + \rho^2_{yz} - 2\rho_{xy}\rho_{yz}\rho_{xz}\right)}{1 - \rho^2_{xz}}\right\} + \theta_1 \frac{\left(\rho_{yx} - \rho_{yz}\rho_{xz}\right)^2}{1 - \rho^2_{xz}}\right]$$

or

$$MSE\left(t_{67(2)}\right) \approx \bar{Y}^2 C_y^2\left[\theta_2\left(1 - \rho^2_{y.xz}\right) + \theta_1 \rho^2_{yx.z}\left(1 - \rho^2_{yz}\right)\right]. \quad (5.3.17)$$

Comparing (5.3.17) with (4.9.12) we see that the estimator proposed by Ahmed et al. (2009) and by Roy (2003) are the same with respect to mean square error. The difference between the two estimators is that Roy's (2003) estimator is a regression-type estimator and the estimator proposed by Ahmed et al. (2009) is a ratio-type estimator. Ahmed et al. (2007) have proposed another estimator by interchanging the roles of the variables. The estimator is:

$$t_{68(2)} = \bar{y}_2\left(\frac{\bar{z}_1}{\bar{z}_2}\right)^{\alpha_1}\left(\frac{\bar{X}}{\bar{x}_1}\right)^{\alpha_2}\left(\frac{\bar{X}}{\bar{x}_2}\right)^{\alpha_3}. \tag{5.3.18}$$

The mean square error of (5.3.18) is the same as given in (4.9.12) with the change of subscripts.

5.3.4 Singh and Upadhyaya (1995) Estimator

Singh and Upadhyaya (1995) proposed a general estimator using the coefficient of variation of one of the auxiliary variables. The proposed estimator is:

$$t_{69(2)} = \bar{y}_2\left(\frac{\bar{x}_1}{\bar{x}_2}\right)\left(\frac{\bar{Z}+C_z}{\bar{z}_1+C_z}\right)^{\alpha_2}, \tag{5.3.19}$$

which reduces to the classical ratio estimator of two-phase sampling for the no information case when $\alpha_2 = 0$. Using notations from section 1.4 in (5.3.19), we have:

$$t_{69(2)} - \bar{Y} = \bar{e}_{y_2} + \frac{\bar{Y}}{\bar{X}}\left(\bar{e}_{x_1} - \bar{e}_{x_2}\right) - \alpha_2\frac{\bar{Y}}{\bar{Z}+C_z}\bar{e}_{z_1}.$$

The mean square error of $t_{69(2)}$ is:

$$MSE\left(t_{69(2)}\right) \approx E\left[\bar{e}_{y_2} + \frac{\bar{Y}}{\bar{X}}\left(\bar{e}_{x_1} - \bar{e}_{x_2}\right) - \alpha_2\frac{\bar{Y}}{\bar{Z}+C_z}\bar{e}_{z_1}\right]^2.$$

Expanding the square and applying expectation, the mean square error is:

$$MSE\left(t_{69(2)}\right) \approx \bar{Y}^2\left[\theta_2 C_y^2 + \alpha_1^2\left(\theta_2 - \theta_1\right)C_x^2 + \alpha_2^2\theta_1\left(\frac{\bar{Z}}{\bar{Z}+C_z}\right)^2 C_z^2\right.$$

$$+2\left(\theta_1-\theta_2\right)C_yC_x\rho_{xy}-2\alpha_2\theta_1\left(\frac{\bar{Z}}{\bar{Z}+C_z}\right)C_yC_z\rho_{yz}\Bigg].$$

(5.3.20)

Optimum value of α_2 which minimizes (5.3.20) is:

$$\alpha_2=\left(\frac{\bar{Z}+C_z}{\bar{Z}}\right)\frac{C_y}{C_z}\rho_{yz}. \tag{5.3.21}$$

Using (5.3.21) in (5.3.20) and simplifying, the mean square error of $t_{69(2)}$ is:

$$MSE\left(t_{69(2)}\right)\approx\bar{Y}^2\left[C_y^2\left(\theta_2-\theta_1\rho_{yz}^2\right)+\left(\theta_2-\theta_1\right)\left\{\left(C_x-C_y\rho_{xy}\right)^2-C_y^2\rho_{xy}^2\right\}\right].$$

(5.3.22)

A modification of $t_{69(2)}$ is readily written by replacing auxiliary variables as:

$$t_{70(2)}=\bar{y}_2\left(\frac{\bar{z}_1}{\bar{z}_2}\right)\left(\frac{\bar{X}+C_x}{\bar{x}_1+C_x}\right)^{\alpha_2}.$$

The mean square error of $t_{70(2)}$ is directly written from (5.3.22) by interchanging subscripts.

5.4 Singh-Upadhyaya-Chandra (2004) Estimators

Singh et al. (2004) have proposed a generalization of estimators proposed by Upadhyaya and Singh (2001) discussed in (5.3.4). We discuss below estimators proposed by Singh et al. (2004).

5.4.1 Singh-Upadhyaya-Chandra (2004) Estimator-I

The first of two estimators proposed by Singh et al. (2004) is:

$$t_{71(2)}=\bar{y}_2\left(\frac{\bar{x}_1}{\bar{x}_2}\right)\left\{\frac{\bar{Z}\beta_2(z)+C_z}{\bar{z}_1\beta_2(z)+C_z}\right\}^{\alpha_2}, \tag{5.4.1}$$

where $\beta_2(z)$ is the coefficient of Kurtosis of auxiliary variable Z. The estimator $t_{71(2)}$ produces the classical ratio estimator of two-phase sampling as a special case for $\alpha_2 = 0$. The Singh and Upadhyaya (2001) estimator, $t_{62(2)}$, emerges as a special case of $t_{71(2)}$ for $\beta_2(z) = 1$. The estimator $t_{46(2)}$ proposed by Upadhyaya and Singh (2001) emerges as a special case of (5.4.1) for $a_1 = p(1-a)$. Substituting $\bar{y}_2 = \bar{Y} + \bar{e}_{y_2}$ etc. in (5.4.1) we have:

$$t_{71(2)} - \bar{Y} = \bar{e}_{y_2} + \frac{\bar{Y}}{\bar{X}}\left(\bar{e}_{x_1} - \bar{e}_{x_2}\right) - \alpha_2 \frac{\bar{Y}\beta_2(z)}{\bar{Z}\beta_2(z) + C_z}\bar{e}_{z_1}.$$

Squaring and applying expectation, the mean square error of $t_{71(2)}$ is:

$$MSE\left(t_{71(2)}\right) \approx E\left[\bar{e}_{y_2} + \frac{\bar{Y}}{\bar{X}}\left(\bar{e}_{x_1} - \bar{e}_{x_2}\right) - \alpha_2 \frac{\bar{Y}\beta_2(z)}{\bar{Z}\beta_2(z) + C_z}\bar{e}_{z_1}\right]^2. \tag{5.4.2}$$

Singh et al. (2004) have shown that the optimum value of α_2 which minimizes (5.4.2) is:

$$\alpha_2 = \left\{\frac{\bar{Z}\beta_2(z)}{\bar{Z}\beta_2(z) + C_z}\right\}^{-1} \frac{C_y}{C_z}\rho_{yz}. \tag{5.4.3}$$

Using (5.4.3) in (5.4.2) and simplifying, the mean square error of $t_{71(2)}$ is:

$$MSE\left(t_{71(2)}\right) \approx \bar{Y}^2\left[C_y^2\left(\theta_2 - \theta_1\rho_{yz}^2\right) + \left(\theta_2 - \theta_1\right)\left\{\left(C_x - C_y\rho_{xy}\right)^2 - C_y^2\rho_{xy}^2\right\}\right]. \tag{5.4.4}$$

Comparison of (5.4.4) with (5.3.22) immediately shows that the Singh and Upadhyaya (1995) estimator and Singh et al. (2004) estimator are in the same equivalence class. A modification of $t_{71(2)}$ may be immediately written from (5.4.1) as:

$$t_{72(2)} = \bar{y}_2\left(\frac{\bar{z}_1}{\bar{z}_2}\right)\left\{\frac{\bar{X}\beta_2(x) + C_x}{\bar{x}_1\beta_2(x) + C_x}\right\}^{\alpha_2}.$$

The mean square error of $t_{72(2)}$ is directly written from (5.4.4) just by interchanging the subscripts.

5.4.2 Singh-Upadhyaya- Chandra (2004) Estimator–II

Singh et al. (2004) have proposed a more general estimator for population mean in two phase sampling using information of two auxiliary variables. The proposed estimator is:

$$t_{73(2)} = \bar{y}_2\left(\frac{\bar{x}_1}{\bar{x}_2}\right)^{\alpha_1}\left(\frac{a\bar{Z}+b}{a\bar{z}_1+b}\right)^{\alpha_2}\left(\frac{a\bar{Z}+b}{a\bar{z}_2+b}\right)^{\alpha_3}. \quad (5.4.5)$$

The estimator $t_{73(2)}$ produces various estimators as special cases for different choices of the constants involved. A simple special case of $t_{73(2)}$ is the classical ratio estimator of two phase sampling which can be obtained for $\alpha_1 = 0$ and $\alpha_2 = 0 = \alpha_3$. Expanding the estimator $t_{73(2)}$ to linear term only, we have:

$$t_{73(2)} - \bar{Y} = \bar{e}_{y_2} + \frac{\bar{Y}}{\bar{X}}\alpha_1\left(\bar{e}_{x_1} - \bar{e}_{x_2}\right) - \bar{Y}\left(\frac{a}{a\bar{Z}+b}\right)\alpha_2\bar{e}_{z_1} - \bar{Y}\left(\frac{a}{a\bar{Z}+b}\right)\alpha_3\bar{e}_{z_2},$$

or

$$t_{73(2)} - \bar{Y} = \bar{e}_{y_2} + \frac{\bar{Y}}{\bar{X}}\alpha_1\left(\bar{e}_{x_1} - \bar{e}_{x_2}\right) - \phi\alpha_2\bar{e}_{z_1} - \phi\alpha_3\bar{e}_{z_2}, \quad (5.4.6)$$

where $\phi = \left(a\bar{Y}\right)/\left(a\bar{Z}+b\right)$. Squaring (5.4.6) and applying expectation, the mean square error of $t_{73(2)}$ is:

$$MSE\left(t_{73(2)}\right) = E\left[\bar{e}_{y_2} + \frac{\bar{Y}}{\bar{X}}\alpha_1\left(\bar{e}_{x_1} - \bar{e}_{x_2}\right) - \bar{Y}\alpha_2\phi\bar{e}_{z_1} - \bar{Y}\alpha_3\phi\bar{e}_{z_2}\right]^2. \quad (5.4.7)$$

Singh et al. (2004) have shown that the estimating equations to obtain values of unknown coefficients which minimize (5.4.7) are:

$$\alpha_1 C_x + \alpha_3\phi\bar{Z}C_z\rho_{xz} - C_y\rho_{xy} = 0, \quad (5.4.8)$$

$$\theta_1\alpha_2\phi\bar{Z}C_z + \theta_1\alpha_3\phi\bar{Z}C_z - \theta_1 C_y\rho_{yz} = 0, \quad (5.4.9)$$

and

$$\alpha_1\left(\theta_2 - \theta_1\right)C_x\rho_{xz} + \alpha_2\phi\theta_1\bar{Z}C_z + \alpha_3\theta_2\phi\bar{Z}C_z - \theta_2 C_y\rho_{yz} = 0. \quad (5.4.10)$$

Solving (5.4.8), (5.4.9) and (5.4.10) simultaneously, we have:

$$\alpha_1 = \frac{C_y}{C_x}\left(\frac{\rho_{xy} - \rho_{yz}\rho_{xz}}{1-\rho_{xz}^2}\right) = \frac{\bar{Y}}{\bar{X}}\beta_{yx.z}, \tag{5.4.11}$$

$$\alpha_2 = \frac{\rho_{xz}C_y}{\bar{Z}C_z}\frac{1}{\phi}\left(\frac{\rho_{yz} - \rho_{xy}\rho_{xz}}{1-\rho_{xz}^2}\right) = \frac{\bar{Y}\rho_{xz}}{\phi}\beta_{yz.x}, \tag{5.4.12}$$

and

$$\alpha_3 = \frac{C_y}{\bar{Z}C_z}\frac{1}{\phi}\left(\frac{\rho_{xy} - \rho_{yz}\rho_{xz}}{1-\rho_{xz}^2}\right) = \frac{\bar{Y}}{\phi}\beta_{yx.z}. \tag{5.4.13}$$

Using (5.4.11), (5.4.12) and (5.4.13) in (5.4.7) and simplifying, the mean square error of Singh et al. (2004) general estimator $t_{73(2)}$ is:

$$MSE\left(t_{73(2)}\right) \approx \bar{Y}^2C_y^2\left[\theta_2\left\{1 - \frac{\left(\rho_{yx}^2 + \rho_{yz}^2 - 2\rho_{xy}\rho_{yz}\rho_{xz}\right)}{1-\rho_{xz}^2}\right\} + \theta_1\frac{\left(\rho_{yx} - \rho_{yz}\rho_{xz}\right)^2}{1-\rho_{xz}^2}\right],$$

or

$$MSE\left(t_{73(2)}\right) \approx \bar{Y}^2C_y^2\left[\theta_2\left(1-\rho_{y.xz}^2\right) + \theta_1\rho_{yx.z}^2\left(1-\rho_{yz}^2\right)\right]. \tag{5.4.14}$$

Comparison of (5.4.14) with (5.3.17) immediately shows that the Singh et al. (2004) estimator is in equivalence class with the Ahmad et al. (2007) estimator. Further, both these estimators are in equivalence class with Roy's (2003) estimator. A modification of $t_{73(2)}$ proposed by Ahmad et al. (2007) is:

$$t_{74(2)} = \bar{y}_2\left(\frac{\bar{z}_1}{\bar{z}_2}\right)^{\alpha_1'}\left(\frac{a\bar{X}+b}{a\bar{x}_1+b}\right)^{\alpha_2'}\left(\frac{a\bar{X}+b}{a\bar{x}_1+b}\right)^{\alpha_3'}.$$

The mean square error of $t_{74(2)}$ may be written immediately:

$$MSE\left(t_{74(2)}\right) \approx \bar{Y}^2C_y^2\left[\theta_2\left(1-\rho_{y.xz}^2\right) + \theta_1\rho_{yz.x}^2\left(1-\rho_{yx}^2\right)\right]. \tag{5.4.15}$$

Comparing (5.4.14) with (5.4.15) we have:

$$MSE\left(t_{73(2)}\right) - MSE\left(t_{74(2)}\right) = \bar{Y}^2C_y^2\left[\theta_1\rho_{yx.z}^2\left(1-\rho_{yz}^2\right) - \theta_1\rho_{yz.x}^2\left(1-\rho_{yx}^2\right)\right]$$

Now using the facts that $\rho_{yx.z}^2\left(1-\rho_{yz}^2\right) = \rho_{y.xz}^2 - \rho_{yz}^2$ and $\rho_{yz.x}^2\left(1-\rho_{yx}^2\right) = \rho_{y.xz}^2 - \rho_{yx}^2$, we have:

$$MSE\left(t_{73(2)}\right)-MSE\left(t_{74(2)}\right)=\theta_1\bar{Y}^2C_y^2\left(\rho_{yx}^2-\rho_{yz}^2\right). \quad (5.4.16)$$

The expression (5.4.16) readily shows that if the squared simple correlation between Y and auxiliary variable X is higher, then the Singh et al. (2004) estimator will perform better as compared with the Ahmad et al. (2007) estimator and vice versa.

Singh and Upadhyaya (1995), Upadhyaya and Singh (2001), Singh et al. (2004) etc. have used the coefficient of variation and/or Coefficient of Skewness $\beta_1(z)$ and/or coefficient of Kurtosis $\beta_2(z)$ (unit less quantities) of auxiliary variables in the construction of estimators without multiplying it with the quantity unit of measurement like mean/standard deviation.

The problem in construction of these types of estimators is that when one is adding only a pure number (unit free quantity like C_z or $\beta_2(z)$ in a quantity having unit of measurement like $\bar{Z}C_z$ or $\bar{Z}\beta_2(z)$, this addition is questionable in the sense that, if someone adds 3 to 4 kg of wheat, then what will be the 7? Either only a pure number 7, or 7 kg of wheat. Obviously, if we cannot decide the unit of measurement of the nominator and denominator, then what about the ratio of being a pure number for the right construction of ratio type estimators?

Also to explain more, suppose $\beta_2(z)=3$ and $C_z=3$. We expect t_7 to be in Kilograms, but it turns out that $\bar{y}_2\left(\frac{\bar{x}_1}{\bar{x}_2}\right)\left(\frac{2\bar{Z}+3}{2\bar{z}_1+3}\right)$ is a pure number, which is not.

5.5 Generalized Regression and Ratio Estimators by Samiuddin and Hanif (2007)

We have discussed number of estimators of two phase sampling in the previous chapter. We have also discussed various generalized estimators of two phase sampling in the preceding sections of this chapter. Samiuddin and Hanif (2007) have discussed different aspects of the use of auxiliary information in two-phase sampling depending upon availability of population information of auxiliary variables. We will discuss all these aspects along these lines.

5.5.1 Estimators for No Information Case (NIC)

What Samiuddin and Hanif (2007) have termed the No Information Case (NIC), is a situation when only sample information is available.

5.5.1.1. Regression Estimator for No Information Case

Samiuddin and Hanif (2007) have proposed ratio and regression estimators of population mean parallel to methods discussed in section 3.4. Samiuddin and Hanif (2007) have argued that when only sample information is available then a regression estimator of population mean in two-phase sampling is:

$$t_{75(2)} = \bar{y}_2 + \alpha\left(\bar{x}_1 - \bar{x}_2\right) + \beta\left(\bar{z}_1 - \bar{z}_2\right). \tag{5.5.1}$$

Using $\bar{y}_2 = \bar{Y} + \bar{e}_{y_2}$ etc. in (5.5.1):

$$t_{75(2)} - \bar{Y} = \bar{e}_{y_2} + \alpha\left(\bar{e}_{x_1} - \bar{e}_{x_2}\right) + \beta\left(\bar{e}_{z_1} - \bar{e}_{z_2}\right).$$

The mean square error of $t_{75(2)}$ is:

$$MSE\left(t_{75(2)}\right) = E\left[\bar{e}_{y_2} + \alpha\left(\bar{e}_{x_1} - \bar{e}_{x_2}\right) + \beta\left(\bar{e}_{z_1} - \bar{e}_{z_2}\right)\right]^2. \tag{5.5.2}$$

The optimum values of α and β which minimizes (5.5.2) are:

$$\alpha = \frac{\bar{Y}C_y\left(\rho_{xy} - \rho_{xz}\rho_{yz}\right)}{\bar{X}C_x\left(1 - \rho_{xz}^2\right)} = \beta_{yx.z}, \tag{5.5.3}$$

and

$$\beta = \frac{\bar{Y}C_y\left(\rho_{yz} - \rho_{xz}\rho_{xy}\right)}{\bar{Z}C_z\left(1 - \rho_{xz}^2\right)} = \beta_{yz.x}. \tag{5.5.4}$$

Substituting values of α and β from (5.5.3) and (5.5.4) in (5.5.2) and simplifying, the mean square error of $t_{75(2)}$ is:

$$MSE\left(t_{75(2)}\right) \approx \bar{Y}^2C_y^2\left[\theta_2\left(1 - \rho_{y.xz}^2\right) + \theta_1\rho_{y.xz}^2\right]. \tag{5.5.5}$$

The mean square error of $t_{75(2)}$, given in (5.5.5), is the same as the mean square error of $t_{57(2)}$ proposed by Hanif et al. (2010) and hence the estimator proposed by Hanif et al. (2010) is in the same equivalence class with the estimator proposed by Samiuddin and Hanif (2007). The estimator $t_{75(2)}$ proposed by Samiuddin and Hanif (2007) is much simpler as compared with the estimator proposed by Hanif et al. (2010).

5.5.1.2. Ratio Estimator for No Information Case

Samiuddin and Hanif (2007) have also proposed a generalized chain ratio-type estimator of population mean using information of two auxiliary variables in no information case. The proposed estimator is:

$$t_{76(2)} = \bar{y}_2\left(\frac{\bar{x}_1}{\bar{x}_2}\right)^p\left(\frac{\bar{z}_1}{\bar{z}_2}\right)^q, \tag{5.5.6}$$

where p and q are constants to be determined. The estimator (5.5.6) is an extension of $t_{62(2)}$, proposed by Srivastava (1971), when two auxiliary variables are used. Substituting $\bar{y}_2 = \bar{Y} + \bar{e}_{y_2}$ etc. in (5.5.6) and expanding up to linear term only, we have from (5.5.6):

$$t_{76(2)} - \bar{Y} = \bar{e}_{y_2} + p\frac{\bar{Y}}{\bar{X}}\left(\bar{e}_{x_1} - \bar{e}_{x_2}\right) + q\frac{\bar{Y}}{\bar{Z}}\left(\bar{e}_{z_1} - \bar{e}_{z_2}\right).$$

The mean square error of $t_{76(2)}$ is:

$$MSE\left(t_{76(2)}\right) = E\left[\bar{e}_{y_2} + p\frac{\bar{Y}}{\bar{X}}\left(\bar{e}_{x_1} - \bar{e}_{x_2}\right) + q\frac{\bar{Y}}{\bar{Z}}\left(\bar{e}_{z_1} - \bar{e}_{z_2}\right)\right]^2. \tag{5.5.7}$$

The optimum choice of p and q, which minimize (5.5.7), is:

$$p = \frac{C_y\left(\rho_{xy} - \rho_{xz}\rho_{yz}\right)}{C_x\left(1-\rho_{xz}^2\right)} = \frac{\bar{Y}}{\bar{X}}\beta_{yx.z}, \tag{5.5.8}$$

and

$$q = \frac{C_y\left(\rho_{yz} - \rho_{xy}\rho_{xz}\right)}{C_z\left(1-\rho_{xz}^2\right)} = \frac{\bar{Y}}{\bar{Z}}\beta_{yz.x}. \tag{5.5.9}$$

Samiuddin and Hanif (2007) have further shown that use of optimum values of p and q in (5.5.7) produces the same mean square error of $t_{76(2)}$ as the mean square error of $t_{75(2)}$ given in (5.5.5) and of $t_{57(2)}$ given in (4.10.11). It is worth mentioning here that the generalized chain ratio estimator and regression estimator proposed by Samiuddin and Hanif

(2007) and regression-in-ratio-product estimator proposed by Hanif et al. (2010) for the no information case are in the same equivalence class.

5.5.2 Estimators for Partial Information Case

We have discussed in section 3.4 and 4.10 that the partial information case is a situation when we have population information for one of two auxiliary variables.

5.5.2.1 Regression Estimator for Partial Information Case

Samiuddin and Hanif (2007) have proposed a regression estimator for the partial information case in two-phase sampling, i.e.:

$$t_{77(2)} = \bar{y}_2 + \alpha_1\left(\bar{x}_1 - \bar{x}_2\right) + \beta_1\left(\bar{z}_1 - \bar{z}_2\right) + \alpha\left(\bar{X} - \bar{x}_2\right). \qquad (5.5.10)$$

Using notation from section 1.4, we have:

$$t_{77(2)} - \bar{Y} = \bar{e}_{y_2} + \alpha_1\left(\bar{e}_{x_1} - \bar{e}_{x_2}\right) + \beta_1\left(\bar{e}_{z_1} - \bar{e}_{z_2}\right) - \alpha\bar{e}_{x_2}.$$

Squaring and applying expectation, the mean square error of $t_{77(2)}$ is:

$$MSE\left(t_{77(2)}\right) = E\left[\bar{e}_{y_2} + \alpha_1\left(\bar{e}_{x_1} - \bar{e}_{x_2}\right) + \beta_1\left(\bar{e}_{z_1} - \bar{e}_{z_2}\right) - \alpha\bar{e}_{x_2}\right]^2. \qquad (5.5.11)$$

The optimum choices of unknowns which minimize (5.5.11) are:

$$\alpha_1 = \frac{\bar{Y}C_y\rho_{zx}\left(\rho_{xy}\rho_{xz} - \rho_{yz}\right)}{\bar{X}C_x\left(1-\rho_{xz}^2\right)} = -\beta_{zx}\beta_{yz.x}, \qquad (5.5.12)$$

$$\beta_1 = \frac{\bar{Y}C_y\left(\rho_{yz} - \rho_{xy}\rho_{xz}\right)}{\bar{X}C_x\left(1-\rho_{xz}^2\right)} = \beta_{yz.x}, \qquad (5.5.13)$$

and

$$\alpha = \frac{\bar{Y}C_y}{\bar{X}C_x}\rho_{xy} = \beta_{yx}. \qquad (5.5.14)$$

Substituting values of α_1, β_1 and α from (5.5.12), (5.5.13) and (5.5.14) in (5.5.11) and simplifying, the mean square error of $t_{77(2)}$ is:

$$MSE\left(t_{77(2)}\right) \approx \bar{Y}^2 C_y^2 \left[\theta_2 \left\{ \frac{1-\rho_{xy}^2-\rho_{xz}^2-\rho_{yz}^2+2\rho_{xy}\rho_{xz}\rho_{yz}}{1-\rho_{xz}^2} \right\} + \theta_1 \left\{ \frac{\left(\rho_{yz}-\rho_{xy}\rho_{yz}\right)^2}{1-\rho_{xz}^2} \right\} \right],$$

or

$$MSE\left(t_{77(2)}\right) \approx \bar{Y}^2 C_y^2 \left[\theta_2 \left(1-\rho_{y.xz}^2\right) + \theta_1 \rho_{yz.x}^2 \left(1-\rho_{xy}^2\right) \right]. \qquad (5.5.15)$$

Immediate comparison of (5.5.15) with (4.9.12) shows that Roy's (2003) estimator and the Samiuddin and Hanif (2007) estimator of the partial information case are in the same equivalence class.

5.5.2.2 Ratio Estimator for Partial Information Case

Samiuddin and Hanif (2007) have also proposed the following generalized chain ratio estimator for the partial information case:

$$t_{78(2)} = \bar{y}_2 \left(\frac{\bar{x}_1}{\bar{x}_2} \right)^{\alpha_1} \left(\frac{\bar{z}_1}{\bar{z}_2} \right)^{\beta_1} \left(\frac{\bar{X}}{\bar{x}_2} \right)^{\alpha}, \qquad (5.5.16)$$

where α_1, β_1 and α are constants to be determined. Substituting values of sample means in terms of population means in (5.5.16) and simplifying, we have:

$$t_{78(2)} - \bar{Y} = \bar{e}_{y_2} + \alpha_1 \frac{\bar{Y}}{\bar{X}}\left(\bar{e}_{x_1} - \bar{e}_{x_2}\right) + \beta_1 \frac{\bar{Y}}{\bar{Z}}\left(\bar{e}_{z_1} - \bar{e}_{z_2}\right) - \frac{\bar{Y}}{\bar{X}} \alpha \bar{e}_{x_2},$$

or $t_{78(2)} - \bar{Y} = \bar{e}_{y_2} + A_1\left(\bar{e}_{x_1} - \bar{e}_{x_2}\right) + B_1\left(\bar{e}_{z_1} - \bar{e}_{z_2}\right) - A\bar{e}_{x_2}$, (5.5.17)

where $A_1 = \left(\bar{Y}/\bar{X}\right)\alpha_1$, $B_1 = \left(\bar{Y}/\bar{Z}\right)\beta_1$ and $A = \left(\bar{Y}/\bar{X}\right)\alpha$. The mean square error of $t_{78(2)}$ is immediately written from (5.5.17) as:

$$MSE\left(t_{78(2)}\right) = E\left[\bar{e}_{y_2} + A_1\left(\bar{e}_{x_1} - \bar{e}_{x_2}\right) + B_1\left(\bar{e}_{z_1} - \bar{e}_{z_2}\right) - A\bar{e}_{x_2}\right]^2. \qquad (5.5.18)$$

Comparing (5.5.18) with (5.5.11) we see that optimum values of unknowns in (5.5.18) are the same as given in (5.5.12), (5.5.13) and (5.5.14). Samiuddin and Hanif (2007) have further shown that the minimum mean square error of $t_{78(2)}$ is the same as the mean square error of Roy's (2003) estimator, given in (4.9.12), and the mean square error of the regression estimator of the partial information case proposed by Samiuddin and Hanif (2007) given in (5.5.15). We, therefore, say that all three of these estimators are in the same equivalence class. We also see that the regression estimator of the partial information case proposed by Samiuddin and Hanif (2007), given in (5.5.10), is the simplest of the three estimators with respect to its applicability.

5.5.3 Estimators for Full Information Case

Samiuddin and Hanif (2007) have termed the full information case when population information of both auxiliary variables is available. Samiuddin and Hanif (2007) have argued that the regression estimator for the full information case, in single phase sampling, always performs better and this is easily seen by looking at (3.4.11). Samiuddin and Hanif (2007) have further argued that availability of population information of auxiliary variables helps a lot in increasing precision of estimates in two-phase sampling. Based on this fact they have proposed a regression and generalized ratio estimator in two-phase sampling when population information of both auxiliary variables is available.

5.5.3.1 Regression Estimator for Full Information Case

The regression estimator, proposed by Samiuddin and Hanif (2007), in the full information case is:

$$t_{79(2)} = \bar{y}_2 + \alpha\left(\bar{X} - \bar{x}_2\right) + \beta\left(\bar{Z} - \bar{z}_2\right), \tag{5.5.19}$$

where α and β are constants to be determined. Utilizing the relation $\bar{y}_2 = \bar{Y} + \bar{e}_{y_2}$ etc. in (5.5.19), we have:

$$t_{79(2)} - \bar{Y} = \bar{e}_{y_2} - \alpha\bar{e}_{x_2} - \beta\bar{e}_{z_2},$$

and hence the mean square error of $t_{79(2)}$ is:

$$MSE\left(t_{79(2)}\right) = E\left[\bar{e}_{y_2} - \alpha\bar{e}_{x_2} - \beta\bar{e}_{z_2}\right]^2. \tag{5.5.20}$$

The optimum values of α and β which minimizes (5.5.20) are:

$$\alpha = \frac{\bar{Y}C_y}{\bar{X}C_x}\left(\frac{\rho_{yx} - \rho_{yz}\rho_{xz}}{1-\rho_{xz}^2}\right) = \beta_{yx.z}; \tag{5.5.21}$$

and

$$\beta = \frac{\bar{Y}C_y}{\bar{X}C_z}\left(\frac{\rho_{yz} - \rho_{xy}\rho_{xz}}{1-\rho_{xz}^2}\right) = \beta_{yz.x}. \tag{5.5.22}$$

Using (5.5.21) and (5.5.22) in (5.5.20) and simplifying, the mean square error of $t_{79(2)}$ is:

$$MSE\left(t_{79(2)}\right) \approx \theta_2 \bar{Y}^2 C_y^2 \left(1-\rho_{y.xz}^2\right). \tag{5.5.23}$$

The mean square error given in (5.5.23) is the same as the mean square error of $t_{61(2)}$ proposed by Hanif et al. (2010), which is given in (4.10.23). So, the estimator $t_{79(2)}$ is in equivalence class with $t_{61(2)}$.

5.5.3.2 Ratio Estimator for Full Information Case

Samiuddin and Hanif (2007) have also proposed a generalized chain ratio estimator for the full information case as:

$$t_{80(2)} = \bar{y}_2 \left(\frac{\bar{X}}{\bar{x}_2}\right)^{\alpha} \left(\frac{\bar{Z}}{\bar{z}_2}\right)^{\beta}. \tag{5.5.24}$$

The estimator $t_{80(2)}$ reduces to $t_{25(2)}$ for $\alpha = \beta = 1$. Using the relation $\bar{y}_2 = \bar{Y} + \bar{e}_{y_2}$ etc., the estimator $t_{80(2)}$ can be put in the form:

$$t_{80(2)} - \bar{Y} = \bar{e}_{y_2} - \alpha\frac{\bar{Y}}{\bar{X}}\bar{e}_{x_2} - \beta\frac{\bar{Y}}{\bar{Z}}\bar{e}_{z_2};$$

or $t_{80(2)} - \bar{Y} = \bar{e}_{y_2} - A\bar{e}_{x_2} - B\bar{e}_{z_2}$;

where $A = \left(\bar{Y}/\bar{X}\right)\alpha$ and $B = \left(\bar{Y}/\bar{Z}\right)\beta$. The mean square error of $t_{80(2)}$ is, therefore:

$$MSE\left(t_{80(2)}\right) = E\left[\bar{e}_{y_2} - A\bar{e}_{x_2} - B\bar{e}_{z_2}\right]^2. \tag{5.5.25}$$

Comparing (5.5.25) with (5.5.20), we see that the optimum values of A and B are the same as given in (5.5.21) and (5.5.22). The mean square error of $t_{80(2)}$ is the same as given in (5.5.23). We, therefore, say that estimators $t_{61(2)}$, $t_{79(2)}$ and $t_{80(2)}$ are in equivalence class.

5.5 Some More Generalized Estimators by Samiuddin and Hanif (2007)

Mohanty (1967) proposed his regression-in-ratio estimator using information of two auxiliary variables. We have discussed the Mohanty (1967) estimator alongside some of its modifications in section 4.3.1. Samiuddin and Hanif (2007) have proposed three generalizations of the Mohanty (1967) estimators under various situations discussed in section 3.4, 4.10 and 5.5 viz, no information case, partial information case and full information case. The core idea in developing these generalized estimators is to use a linear combination of two separate Mohanty (1967) estimators under different situations. The base estimator selected by Samiuddin and Hanif (2007) was of the form:

$$t = at_1 + (1-a)t_2, \qquad (5.6.1)$$

where t_1 and t_2 are appropriate Mohanty (1967) estimators.

5.6.1 Generalized Estimator for No Information Case

The first of three generalized estimators proposed by Samiuddin and Hanif (2007) for the no information case is:

$$t_{81(2)} = at_1 + (1-a)t_2, \qquad (5.6.2)$$

where

$$t_1 = \left[\bar{y}_2 + k_1(\bar{x}_1 - \bar{x}_2)\right]\frac{\bar{z}_1}{\bar{z}_2},$$

and

$$t_2 = \left[\bar{y}_2 + k_2(\bar{z}_1 - \bar{z}_2)\right]\frac{\bar{x}_1}{\bar{x}_2}.$$

Using notations from section 1.4 in (5.6.2) and simplifying we have:

$$t_{81(2)} - \bar{Y} = \bar{e}_{y_2} + \left[ak_1(1-a)\frac{\bar{Y}}{\bar{X}}\right]\left(\bar{e}_{x_1} - \bar{e}_{x_2}\right) + \left[a\frac{\bar{Y}}{\bar{Z}} + (1-a)k_2\right]\left(\bar{e}_{z_1} - \bar{e}_{z_2}\right),$$

or $$t_{81(2)} - \bar{Y} = \bar{e}_{y_2} + A_1\left(\bar{e}_{x_1} - \bar{e}_{x_2}\right) + A_2\left(\bar{e}_{z_1} - \bar{e}_{z_2}\right), \qquad (5.6.3)$$

where $A_1 = ak_1(1-a)\left(\bar{Y}/\bar{X}\right)$ and $A_2 = a\left(\bar{Y}/\bar{Z}\right) + (1-a)k_2$.

Squaring (5.6.3) and applying expectation, the mean square error of $t_{81(2)}$ is:

$$MSE\left(t_{81(2)}\right) \approx E\left[\bar{e}_{y_2} + A_1\left(\bar{e}_{x_1} - \bar{e}_{x_2}\right) + A_2\left(\bar{e}_{z_1} - \bar{e}_{z_2}\right)\right]^2. \quad (5.6.4)$$

Comparison of (5.6.4) with (5.5.2) immediately shows that estimators $t_{75(2)}$ and $t_{81(2)}$ are in equivalence class. The mean square error of $t_{81(2)}$ is the same as the mean square error of $t_{75(2)}$ given in (5.5.5).

5.6.2 Generalized Estimator for Partial Information Case

The second estimator proposed by Samiuddin and Hanif (2007) for the partial information case is:

$$t_{82(2)} = at_1 + (1-a)t_2. \quad (5.6.5)$$

where

$$t_1 = \left[\bar{y}_2 + k_1\left(\bar{x}_1 - \bar{x}_2\right)\right]\frac{\bar{z}_1}{\bar{z}_2}, \quad (5.6.6)$$

and

$$t_2 = \left[\bar{y}_2 + k_2\left(\bar{z}_1 - \bar{z}_2\right)\right]\frac{\bar{X}}{\bar{x}_1}. \quad (5.6.7)$$

Putting $\bar{y}_2 = \bar{Y} + \bar{e}_{y_2}$ etc. (5.6.6) and simplifying we have:

$$t_1 - \bar{Y} = \bar{e}_{y_2} + k_1\left(\bar{e}_{x_1} - \bar{e}_{x_2}\right) + \frac{\bar{Y}}{\bar{Z}}\left(\bar{e}_{z_1} - \bar{e}_{z_2}\right). \quad (5.6.8)$$

Similarly we have:

$$t_2 - \bar{Y} = \bar{e}_{y_2} + k_2\left(\bar{e}_{z_1} - \bar{e}_{z_2}\right) - \frac{\bar{Y}}{\bar{X}}\left(\bar{e}_{x_1} - \bar{e}_{x_2}\right) - \frac{\bar{Y}}{\bar{X}}\bar{e}_{x_2}. \quad (5.6.9)$$

Using (5.6.8) and (5.6.9) in (5.6.5) and simplifying, we have:

$$t_{82(2)} - \bar{Y} = \bar{e}_{y_2} + A\left(\bar{e}_{x_1} - \bar{e}_{x_2}\right) + B\left(\bar{e}_{z_1} - \bar{e}_{z_2}\right) - C\bar{e}_{x_2},$$

where $A = ak_1 - (1-a)\left(\bar{Y}/\bar{X}\right)$, $B = a\left(\bar{Y}/\bar{Z}\right) + (1-a)k_2$ and $C = (1-a)\left(\bar{Y}/\bar{X}\right)$. The mean square error of $t_{82(2)}$ is, therefore:

$$MSE\left(t_{82(2)}\right) = E\left[\bar{e}_{y_2} + A\left(\bar{e}_{x_1} - \bar{e}_{x_2}\right) + B\left(\bar{e}_{z_1} - \bar{e}_{z_2}\right) - C\bar{e}_{x_2}\right]^2. \quad (5.6.10)$$

Comparing (5.6.10) with (5.5.11), we see that the estimators $t_{77(2)}$ and $t_{82(2)}$ are in the same equivalence class. The mean square error of $t_{82(2)}$ is the same as the mean square error of $t_{77(2)}$, given in (5.5.15) and the mean square error of the Roy (2003) estimator given in (4.9.12).

5.6.3 Generalized Estimator for Full Information Case

Samiuddin and Hanif (2007) proposed a third estimator for the full information case as:

$$t_{83(2)} = at_1 + (1-a)t_2 . \tag{5.6.11}$$

where

$$t_1 = \left[\bar{y}_2 + k_1\left(\bar{x}_1 - \bar{x}_2\right)\right]\left(\frac{\bar{Z}}{\bar{z}_2}\right)^{\gamma}; \tag{5.6.12}$$

and

$$t_2 = \left[\bar{y}_2 + k_2\left(\bar{z}_1 - \bar{z}_2\right)\right]\left(\frac{\bar{X}}{\bar{x}_2}\right)^{\delta}. \tag{5.6.13}$$

Using $\bar{y}_2 = \bar{Y} + \bar{e}_{y_2}$ etc. in (5.6.11) and simplifying, we have:

$$t_1 - \bar{Y} = \bar{e}_{y_2} + k_1\left(\bar{e}_{x_1} - \bar{e}_{x_2}\right) - \gamma\left(\bar{Y}/\bar{Z}\right)\bar{e}_{z_2} . \tag{5.6.14}$$

Similarly we have:

$$t_2 - \bar{Y} = \bar{e}_{y_2} + k_2\left(\bar{e}_{z_1} - \bar{e}_{z_2}\right) - \delta\left(\bar{Y}/\bar{X}\right)\bar{e}_{x_2} . \tag{5.6.15}$$

Substituting (5.6.14) and (5.6.15) in (5.6.11) and simplifying, we have:

$$t_{83(2)} - \bar{Y} = \bar{e}_{y_2} + \alpha_1\left(\bar{e}_{x_1} - \bar{e}_{x_2}\right) + \beta_1\left(\bar{e}_{z_1} - \bar{e}_{z_2}\right) - \alpha\bar{e}_{x_2} - \beta\bar{e}_{z_2} ,$$

where $\alpha_1 = ak_1$, $\beta_1 = (1-a)k_2$, $\alpha = p\gamma\left(\bar{Y}/\bar{Z}\right)$ and $\beta = (1-p)\delta\left(\bar{Y}/\bar{X}\right)$. Squaring and applying expectation, the mean square error of $t_{83(2)}$ is:

$$MSE\left(t_{83(2)}\right) = E\left[\bar{e}_{y_2} + \alpha_1\left(\bar{e}_{x_1} - \bar{e}_{x_2}\right) + \beta_1\left(\bar{e}_{z_1} - \bar{e}_{z_2}\right) - \alpha\bar{e}_{x_2} - \beta\bar{e}_{z_2}\right]^2 . \tag{5.6.16}$$

The optimum choice of unknowns which minimizes (5.6.16) are:

$$\alpha_1 = 0 = \beta_1; \tag{5.6.17}$$

$$\alpha = \frac{\bar{Y}C_y}{\bar{X}C_x}\left(\frac{\rho_{yx} - \rho_{yz}\rho_{xz}}{1-\rho_{xz}^2}\right) = \beta_{yx.z}, \tag{5.6.18}$$

and

$$\beta = \frac{\bar{Y}C_y}{\bar{X}C_z}\left(\frac{\rho_{yz} - \rho_{xy}\rho_{xz}}{1-\rho_{xz}^2}\right) = \beta_{yz.x}. \tag{5.6.19}$$

Substituting the values of constant and simplifying the mean square error of $t_{83(2)}$ is:

$$MSE\left(t_{83(2)}\right) \approx \theta_2 \bar{Y}^2 C_y^2\left(1-\rho_{y.xz}^2\right). \tag{5.6.20}$$

The mean square error of $t_{83(2)}$, given in (5.6.20), is the same as the mean square error of $t_{61(2)}$, given in (4.10.23) and mean square error of $t_{79(2)}$, given in (5.5.23). We therefore conclude that estimators $t_{61(2)}$, $t_{79(2)}$ and $t_{83(2)}$ are in the same equivalence class.

5.6.4 Singh et al. (2008) estimator

Singh et al. (2008) proposed an estimator:

$$t_{84(2)} = w_0\bar{y} + w_1 t_{rd} + w_2 t_{rs}\ , \tag{5.6.21}$$

where

$$t_{rd} = \bar{y}_2\left(\frac{a\bar{x}_1 + b}{a\bar{x}_2 + b}\right)$$

and

$$t_{rs} = \bar{y}_2 \exp\left(\frac{(a\bar{x}_1 + b) - (a\bar{x}_2 + b)}{(a\bar{x}_1 + b) + (a\bar{x}_2 + b)}\right).$$

For the mean square error of $t_{84(2)}$, we proceed as follows:

Taking

$$t_{rd} = \bar{y}_2\left(\frac{a\bar{x}_1 + b}{a\bar{x}_2 + b}\right)$$

Expressing t_{rd} in terms of e's

$$t_{rd} = \bar{Y}(1+e_0)\left[\left(1+\frac{a\bar{X}e_2}{a\bar{X}+b}\right)^{-1} + \left(\frac{a\bar{X}+b}{a\bar{X}e_1}+\frac{a\bar{X}e_2}{a\bar{X}e_1}\right)^{-1}\right].$$

If $\lambda = \dfrac{a\bar{X}}{a\bar{X}+b}$, then

$$t_{rd} = \bar{Y}(1+e_0)\left[(1-\lambda e_2)+\lambda e_1(1-\lambda e_2)\right]$$

or

$$t_{rd} = \bar{Y}(1+e_0)\left[(1+\lambda e_1 - \lambda e_2\right] \tag{5.6.22}$$

Now, taking

$$t_{rs} = \bar{y}_2 \exp\left(\frac{(a\bar{x}_1+b)-(a\bar{x}_2+b)}{(a\bar{x}_1+b)+(a\bar{x}_2+b)}\right)$$

Expressing t_{rs} in terms of e's

$$t_{rd} = \bar{Y}(1+e_0)\exp\left[\left(1+\frac{e_2}{e_1}+\frac{2(a\bar{X}+b)}{a\bar{X}e_1}\right)^{-1} - \left(1+\frac{e_1}{e_2}+\frac{2(a\bar{X}+b)}{a\bar{X}e_2}\right)^{-1}\right]$$

$$t_{rd} = \bar{Y}(1+e_0)\left[(1+\frac{\lambda e_1}{2}-\frac{\lambda e_2}{2}\right] \tag{5.6.23}$$

Using equation (5.6.22) and (5.6.23) in equation (5.6.21)

$$t_{84(2)} = w_0\bar{Y}(1+e_0) + w_1\bar{Y}(1+e_0)\left[(1+\lambda e_1-\lambda e_2\right] + w_2\bar{Y}(1+e_0)\left[(1+\frac{\lambda e_1}{2}-\frac{\lambda e_2}{2}\right]$$

or

$$t_{84(2)} = \bar{Y}(1+e_0)\left[w_0 + w_1(1+\lambda e_1-\lambda e_2)+w_2\left(1+\frac{\lambda e_1}{2}-\frac{\lambda e_2}{2}\right)\right]$$

If $w = \left(w_1 + \dfrac{w_2}{2}\right)$

$$t_{84(2)} = \bar{Y}(1+e_0)\left[1+\lambda e_1 w - \lambda e_2 w\right]$$

$$t_{84(2)} - \bar{Y} = \bar{Y}\left[e_0 + \lambda e_1 w - \lambda e_2 w\right]$$

Squaring, taking expectations and by simplification, we get:

$$MSE\left(t_{84(2)}\right)=\bar{Y}^2\left[\theta_2 C_y^2+(\theta_2-\theta_1)\lambda^2 w^2 C_x^2-2(\theta_2-\theta_1)\lambda^2 w^2 C_y C_x \rho_{xy}\right]$$

Some members of the family of estimators are given in the following table.

a	b	w_0 (q_0)	w_1 (q_1)	w_2 (q_2)	Ratio Estimator (corresponding to w_i for i=0,1,2)	Product Estimator (corresponding to w_i for i=0,1,2)
0	0	1	0	0	$\bar{y}$, the mean per unit estimator	$\bar{y}$, The mean per unit estimator
1	0	0	1	0	$\frac{\bar{y}}{\bar{x}}\bar{X}$, The usual ratio estimator	$\frac{\bar{y}}{\bar{X}}\bar{x}$, The usual product estimator
1	C_x	0	1	0	$\bar{y}\frac{\bar{X}+C_x}{\bar{x}+C_x}$, Sisodia and Dwivedi (1981) estimator	$\bar{y}\frac{\bar{x}+C_x}{\bar{X}+C_x}$, Pandey and Dubey (1988) estimator
1	$\beta_2(x)$	0	1	0	$\bar{y}\frac{\bar{X}+\beta_2(x)}{\bar{x}+\beta_2(x)}$, Singh et. al. (2004) estimator	$\bar{y}\frac{\bar{x}+\beta_2(x)}{\bar{X}+\beta_2(x)}$, Singh et. al. (2004) estimator
$\beta_2(x)$	C_x	0	1	0	$\bar{y}\frac{\bar{X}\beta_2(x)+C_x}{\bar{x}\beta_2(x)+C_x}$, Upadhyaya and Singh (1999) Estimator	$\bar{y}\frac{\bar{x}\beta_2(x)+C_x}{\bar{X}\beta_2(x)+C_x}$, Upadhyaya and Singh (1999) estimator
C_x	$\beta_2(x)$	0	1	0	$\bar{y}\frac{\bar{X}C_x+\beta_2(x)}{\bar{x}C_x+\beta_2(x)}$, Upadhyaya and Singh (1999) Estimator	$\bar{y}\frac{\bar{x}C_x+\beta_2(x)}{\bar{X}C_x+\beta_2(x)}$, Upadhyaya and Singh (1999) estimator

a	b	w_0 (q_0)	w_1 (q_1)	w_2 (q_2)	Ratio Estimator (corresponding to w_i for i=0,1,2)	Product Estimator (corresponding to w_i for i=0,1,2)
1	ρ_{xy}	0	1	0	$\bar{y}\dfrac{\bar{X}+\rho_{xy}}{\bar{x}+\rho_{xy}}$, Singh and Tailor (2003) Estimator	$\bar{y}\dfrac{\bar{x}+\rho_{xy}}{\bar{X}+\rho_{xy}}$, Singh and Tailor (2003) estimator
1	0	0	0	1	$\bar{y}\exp\dfrac{\bar{X}-\bar{x}}{\bar{X}+\bar{x}}$, Bahl and Tuteja (1991) Estimator	$\bar{y}\exp\dfrac{\bar{x}-\bar{X}}{\bar{x}+\bar{X}}$, Bahl and Tuteja (1991) estimator
1	$\beta_2(x)$	0	0	1	$\bar{y}\exp\left[\dfrac{\bar{X}-\bar{x}}{\bar{X}+\bar{x}+2\beta_2(x)}\right]$, Singh et. al. (2007) estimator	$\bar{y}\exp\left[\dfrac{\bar{X}-\bar{x}}{\bar{X}+\bar{x}+2\beta_2(x)}\right]$, Singh et. al. (2007) estimator
1	C_x	0	0	1	$\bar{y}\exp\left[\dfrac{\bar{X}-\bar{x}}{\bar{X}+\bar{x}+2C_x}\right]$, Singh et. al. (2007) estimator	$\bar{y}\exp\left[\dfrac{\bar{x}-\bar{X}}{\bar{X}+\bar{x}+2C_x}\right]$, Singh et. al. (2007) estimator
1	ρ_{xy}	0	0	1	$\bar{y}\exp\left[\dfrac{\bar{X}-\bar{x}}{\bar{X}+\bar{x}+2\rho_{xy}}\right]$, Singh et. al. (2007) estimator	$\bar{y}\exp\left[\dfrac{\bar{x}-\bar{X}}{\bar{X}+\bar{x}+2\rho_{xy}}\right]$, Singh et. al. (2007) estimator

a	b	w_0 (q_0)	w_1 (q_1)	w_2 (q_2)	Ratio Estimator (corresponding to w_i for i=0,1,2)	Product Estimator (corresponding to w_i for i=0,1,2)
$\beta_2(x)$	C_x	0	0	1	$\bar{y}\exp\left[\frac{\beta_2(x)(\bar{X}-\bar{x})}{\beta_2(x)(\bar{X}+\bar{x})+2C_x}\right]$, Singh et. al. (2007) estimator	$\bar{y}\exp\left[\frac{\beta_2(x)(\bar{x}-\bar{X})}{\beta_2(x)(\bar{X}+\bar{x})+2C_x}\right]$, Singh et. al. (2007) estimator
C_x	$\beta_2(x)$	0	0	1	$\bar{y}\exp\left[\frac{C_x(\bar{X}-\bar{x})}{C_x(\bar{X}+\bar{x})+2\beta_2(x)}\right]$, Singh et. al. (2007) estimator	$\bar{y}\exp\left[\frac{C_x(\bar{x}-\bar{X})}{C_x(\bar{X}+\bar{x})+2\beta_2(x)}\right]$, Singh et. al. (2007) estimator
C_x	ρ_{xy}	0	0	1	$\bar{y}\exp\left[\frac{C_x(\bar{X}-\bar{x})}{C_x(\bar{X}+\bar{x})+2\rho_{xy}}\right]$, Singh et. al. (2007) estimator	$\bar{y}\exp\left[\frac{C_x(\bar{x}-\bar{X})}{C_x(\bar{X}+\bar{x})+2\rho_{xy}}\right]$, Singh et. al. (2007) estimator
ρ_{xy}	C_x	0	0	1	$\bar{y}\exp\left[\frac{\rho_{xy}(\bar{X}-\bar{x})}{\rho_{xy}(\bar{X}+\bar{x})+2C_x}\right]$, Singh et. al. (2007) estimator	$\bar{y}\exp\left[\frac{\rho_{xy}(\bar{x}-\bar{X})}{\rho_{xy}(\bar{X}+\bar{x})+2C_x}\right]$, Singh et. al. (2007) estimator
$\beta_2(x)$	ρ_{xy}	0	0	1	$\bar{y}\exp\left[\frac{\rho_{xy}(\bar{X}-\bar{x})}{\rho_{xy}(\bar{X}+\bar{x})+2\beta_2(x)}\right]$, Singh et. al. (2007) estimator	$\bar{y}\exp\left[\frac{\rho_{xy}(\bar{x}-\bar{X})}{\rho_{xy}(\bar{X}+\bar{x})+2\beta_2(x)}\right]$, Singh et. al. (2007) estimator

a	b	w_0 (q_0)	w_1 (q_1)	w_2 (q_2)	Ratio Estimator (corresponding to w_i for i=0,1,2)	Product Estimator (corresponding to w_i for i=0,1,2)
ρ_{xy}	$\beta_2(x)$	0	0	1	$\bar{y}\exp\left[\frac{\beta_2(x)(\bar{X}-\bar{x})}{\beta_2(x)(\bar{X}+\bar{x})+2\rho_{xy}}\right]$, Singh et. al. (2007) estimator	$\bar{y}\exp\left[\frac{\beta_2(x)(\bar{x}-\bar{X})}{\beta_2(x)(\bar{X}+\bar{x})+2\rho_{xy}}\right]$, Singh et. al. (2007) estimator

SECTION-III:

SINGLE AND TWO PHASE SAMPLING WITH MULTI-AUXILIARY VARIABLE

CHAPTER SIX

SINGLE AND TWO PHASE SAMPLING WITH MULTIPLE AUXILIARY VARIABLES

6.1 Introduction

In chapters 4 and 5 we have discussed several estimators using of one and two auxiliary variables under two-phase sampling for estimating the mean of the study variable. The idea of the inclusion of more auxiliary variables in the estimation procedure emerges as more auxiliary variables are likely to bring more precision to estimates. Various estimators have been proposed in the literature which are based upon multiple auxiliary variables and have proved to be more precise as compared with estimators based upon one or two auxiliary variables. The decision of including more auxiliary variables in the estimation process is not merely based upon obtaining high precision but is also based upon affordable cost. Survey statisticians, therefore, have proposed more estimators based upon single or two auxiliary variables rather than estimators based upon multiple auxiliary variables. Use of multiple auxiliary variables in survey sampling has been discussed by a number of researchers. The classical work in this area is, perhaps, of Olkin (1958) who proposed a ratio estimator based on several auxiliary variables. In this Chapter we will discuss some estimators for single and two-phase sampling which are based upon multiple auxiliary variables. We first discuss estimators for single-phase sampling and then present estimators for two phase sampling.

6.2 Estimators for Single Phase Sampling

We have discussed ratio and regression type estimators for single-phase sampling based upon single and two auxiliary variables in Chapters 2 and 3. The estimators discussed in those Chapters may be extended to the case of several auxiliary variables. Various survey statisticians have proposed different estimators for single-phase sampling using information of multiple auxiliary variables. We discuss some of these estimators in the following.

6.2.1 Regression Estimator

The regression estimator for single-phase sampling based upon a single auxiliary variable has been discussed in section 2.2.4. The extension of the regression estimator using multiple auxiliary variables is fairly straightforward and the estimator in this case is given as:

$$t_{m1(1)} = \bar{y} + \sum_{i=1}^{p} \alpha_i \left(\bar{X}_i - \bar{x}_{(i)} \right), \tag{6.2.1}$$

where α_i are constants to be determined so that the mean square error of (6.2.1) is minimum. The vector representation of (6.2.1) is:

$$t_{m1(1)} = \bar{y} + \alpha' \left(\bar{\mathrm{X}} - \bar{\mathrm{x}} \right). \tag{6.2.2}$$

Making the substitution $\bar{y} = \bar{Y} + \bar{e}_y$ and $\bar{\mathrm{x}} = \bar{\mathrm{X}} + \bar{\mathrm{e}}_x$, we can write (6.2.2) as:

$$t_{m1(1)} - \bar{Y} = \bar{e}_y - \alpha' \bar{\mathrm{e}}_x .$$

Squaring the above equation we will have:

$$\left(t_{m1(1)} - \bar{Y} \right)^2 = \bar{e}_y^2 + \alpha' \bar{\mathrm{e}}_x \bar{\mathrm{e}}_x' \alpha - 2\alpha' \bar{e}_y \bar{\mathrm{e}}_x .$$

Applying expectation and using (1.4.4), the mean square error of $t_{m1(1)}$ is:

$$MSE\left(t_{m1(1)} \right) \approx \theta_1 S_y^2 + \theta_1 \alpha' \mathrm{S}_x \alpha - 2\theta_1 \alpha' \mathrm{s}_{yx} . \tag{6.2.3}$$

Differentiating (6.2.3) *w.r.t.* α and equating to zero, the optimum value of α which minimizes (6.2.3) is:

$$\alpha = \mathrm{S}_x^{-1} \mathrm{s}_{yx} , \tag{6.2.4}$$

which is a vector of partial regression coefficients of Y with each auxiliary variable. Using (6.2.4) in (6.2.3) and simplifying we will have:

$$MSE\left(t_{m1(1)} \right) = \theta_1 S_y^2 \left(1 - \mathrm{s}_{yx}' \mathrm{S}_x^{-1} \mathrm{s}_{yx} / S_y^2 \right).$$

Now using $\rho_{y.\mathrm{x}}^2 = \mathrm{s}_{yx}' \mathrm{S}_x^{-1} \mathrm{s}_{yx} / S_y^2$, where $\rho_{y.\mathrm{x}}^2$ is squared multiple correlation between Y and combined effect of all auxiliary variables, the mean square error of $t_{m1(1)}$ is:

$$MSE\left(t_{m1(1)} \right) \approx \theta_1 \bar{Y}^2 C_y^2 \left(1 - \rho_{y.\mathrm{x}}^2 \right). \tag{6.2.5}$$

The mean square error given in (6.2.5) is an extension of the mean square error of the classical regression estimator with single auxiliary variable given in (2.2.12).

6.2.2 Olkin (1958) Ratio Estimator

Olkin (1958) extended the classical ratio estimator to the case of several auxiliary variables when multiple regression of Y on all auxiliary variables passes through the origin. The estimator proposed by Olkin (1958) is:

$$t_{m2(1)} = \bar{y}\sum_{i=1}^{q} W_i \left(\bar{X}_i/\bar{x}_i\right) = \sum_{i=1}^{q} W_i t_{1(1)i} , \tag{6.2.6}$$

where W_i are constants to be determined, under the constraint that $\sum_1^q W_i = 1$, so that the mean square error of $t_{m2(1)}$ is minimum and $t_{1(1)i}$ is the classical ratio estimator using the ith auxiliary variable. The estimator $t_{m2(1)}$ is actually a weighted average of several classical ratio estimators. Defining **V** as a $(q \times q)$ matrix whose ith entry is $MSE\left(t_{1(1)i}\right)$ and (i,j)th entry is $Cov\left(t_{1(1)i}, t_{1(1)j}\right)$. The optimum value of W_i's from (6.2.6) was $W_i = \Sigma_i/\Sigma$, where Σ_i is sum of elements in the ith column of V^{-1} and Σ is sum of all entries in V^{-1}. The minimum variance of $t_{m2(1)}$ derived by Olkin (1958) was:

$$MSE\left(t_{m2(1)}\right) = 1/\Sigma , \tag{6.2.7}$$

where Σ is defined earlier.

6.2.3 Rao and Mudholkar (1967) Ratio-Product Estimator

The ratio estimator proposed by Olkin (1958) is useful when all auxiliary variables have a strong positive correlation with the variable under study. This strong assumption is sometime very difficult meet, and in practice some auxiliary variables have a positive correlation with the variable under study and some have a negative correlation. Rao and Mudholkar (1967) extended the Olkin (1958) estimator further by suggesting the following ratio–product estimator using information of several auxiliary variables:

$$t_{m3(1)} = \bar{y}\left[\sum_{1}^{r} W_i\left(\bar{X}_i/\bar{x}_i\right) + \sum_{r+1}^{r+s} W_i\left(\bar{x}_i/\bar{X}_i\right)\right], \bar{X}_i \neq 0, \tag{6.2.8}$$

where W_i are weights so that $\Sigma W_i = 1$. The estimator (6.2.8) is a combination of two estimators; the first part being Olkin (1958) estimator. Rao and Mudholkar (1967) have shown that the mean square error of (6.2.8) is:

$$MSE\left(t_{m_1}\right) = \theta \bar{Y}^2 \sum_{1}^{r+s} W_i W_j d_{ij} = \theta \bar{Y}^2 \mathrm{W}^{/}\mathrm{DW}, \tag{6.2.9}$$

where **W** is an $(r+s)$ column vector containing weights W_i, **D** is an $(r+s)\times(r+s)$ matrix with entries given as:

$$d_{ij} = C_y^2 - \rho_{yx_i} C_y C_{x_i} + \rho_{yx_j} C_y C_{x_j} + \rho_{ij} C_{x_i} C_{x_j}.$$

The estimator (6.2.8) has much wider applicability as compared with the Olkin (1958) estimator.

6.2.4 Srivastava's (1971) Estimator

Srivastava (1971) proposed a general class of estimators of population mean in single-phase sampling using information of multiple auxiliary variables. The estimator proposed by Srivastava (1971) is based upon a general function $h\left(x_1, x_2, ..., x_q\right)$ of auxiliary variables so that $h(\mathrm{e}) = 1$ and it satisfies the following conditions:

1) Whatever the sample chosen, h assumes values in a bounded, closed convex subset, D, of the p-dimensional real space containing the point θ.
2) In D, the function $h(\mathrm{x})$ is continuous and bounded.
3) The first and second partial derivatives, of $h(\mathrm{x})$ exist and are continuous and bounded in D.

Observing the above properties of $h(\mathrm{x})$, Srivastava (1971) suggested the following estimator of population mean:

$$t_{m4(1)} = \bar{y}\, h(\mathrm{x}). \tag{6.2.10}$$

Srivastava (1971) has shown that the estimator (6.2.10) is unbiased to $O\left(n^{-1}\right)$ by expanding it using Taylor series expansion as:

$$t_{m4(1)} = \bar{y}\left\{h(\mathrm{e}) + (\mathrm{x}-\mathrm{e})' \mathrm{h}'(\mathrm{e}) + (1/2)(\mathrm{x}-\mathrm{e})' \mathrm{H}''\left(\mathrm{x}^*\right)(\mathrm{x}-\mathrm{e})\right\},$$
$$= \bar{Y}\left(1+\xi_0\right)\left\{1+\varepsilon'\mathrm{h}'(\mathrm{e}) + (1/2)\varepsilon'\mathrm{H}''\left(\mathrm{x}^*\right)\varepsilon\right\}, \tag{6.2.11}$$

where $\mathrm{h}'(\mathrm{e})$ is a vector of partial derivatives of $\mathrm{h}(\mathrm{x})$ at point **e**, $H''\left(x^*\right)$ is a matrix of second order derivatives of $\mathrm{h}(\mathrm{x})$ at $\mathbf{x}^*$ with $\mathrm{x}^* = \mathrm{e} + \zeta\varepsilon$ and $0 < \zeta < 1$. Applying expectation on (6.2.11) it may be immediately seen that $t_{m4(1)}$ is unbiased for $\bar{Y}$. Srivastava (1971) has further shown that the mean square error of $t_{m4(1)}$ is:

$$MSE\left(t_{m4(1)}\right) = \theta_1 \bar{Y}^2 C_y^2\left(1-\rho_{y.\mathrm{x}}^2\right), \tag{6.2.12}$$

which is the same as the mean square error of a regression estimator with multiple auxiliary variables. The expression (6.2.12) leads us to the conclusion that any member function $h(\mathrm{x})$ of the Srivastava (1971) estimator will have the same mean square error.

We have seen in Chapters 4 and 5 how auxiliary information helps us in increasing precision of estimates in two-phase sampling. We have also seen in those chapters, estimators based upon two auxiliary variables generally provide more precise results as compared with estimates based upon a smaller number of auxiliary variables. Further we have also seen in Chapter 5 that multiple auxiliary variables may be used in different scenarios in two-phase sampling, which have been named as Full Information Case (FIC), Partial Information Case (PIC), and No Information Case (NIC) by Samiuddin and Hanif (2007). The precision of estimates may further be increased by the inclusion of a larger number of auxiliary variables. In the following sections we will discuss some popular two-phase sampling estimators under the above mentioned scenarios which are based upon multiple auxiliary variables.

6.3 Estimators for Two-Phase Sampling

Various survey statisticians have proposed various estimators for two-phase sampling using information of single or two auxiliary variables. An increase in the number of auxiliary variables is supposed to increase the precision of estimates, as we have seen in the case of the regression estimator in two-phase sampling using single and two auxiliary variables.

We now discuss some popular two-phase sampling estimators which are based upon multiple auxiliary variables.

6.3.1 Regression Estimators

We have discussed the regression estimator for single-phase sampling using information of multiple auxiliary variables in section 6.2.1. We have seen that the estimator is a simple extension of the classical regression estimator with a single auxiliary variable, given in (2.2.10). The extension of the regression estimator for multiple auxiliary variables in two-phase sampling may be constructed analogously using the ideas of section (4.2.3). We now discuss the regression estimator in two-phase sampling using multiple auxiliary variables under different possibilities of availability of auxiliary information.

6.3.1.1 Regression Estimator for No Information Case

The classical regression estimator in two-phase sampling when population information of none of the auxiliary variables is available (NIC) is:

$$t_{m1(2)} = \bar{y}_2 + \sum_{i=1}^{q} \alpha_i \left(\bar{x}_{(i)1} - \bar{x}_{(i)2}\right) = \bar{y}_2 + \alpha' \left(\bar{\mathrm{x}}_1 - \bar{\mathrm{x}}_2\right). \quad (6.3.1)$$

We now derive the mean square error of $t_{m1(2)}$ by using $\bar{\mathrm{x}}_1 = \bar{\mathrm{X}} + \bar{\mathrm{e}}_{x_1}$, $\bar{\mathrm{x}}_2 = \bar{\mathrm{X}} + \bar{\mathrm{e}}_{x_2}$ and $\bar{y}_2 = \bar{Y} + \bar{e}_{y_2}$ in (6.3.1). Making this transformation, we have:

$$t_{m1(2)} - \bar{Y} = \bar{e}_{y_2} + \alpha' \left(\bar{\mathrm{e}}_{x_1} - \bar{\mathrm{e}}_{x_2}\right).$$

Squaring, applying expectation and using (1.4.4), the mean square error of $t_{m1(2)}$ is:

$$MSE\left(t_{m1(2)}\right) = \theta_2 S_y^2 + \left(\theta_2 - \theta_1\right)\alpha' \mathrm{S}_x \alpha - 2\left(\theta_2 - \theta_1\right)\alpha' \mathrm{s}_{yx}. \quad (6.3.2)$$

Differentiating (6.3.2) *w.r.t.* α and equating to zero, the optimum value of α for which the mean square error of $t_{m1(2)}$ is minimized is the same as given in (6.2.4). Further, by using (6.2.4) in (6.3.2) and simplifying we have:

$$MSE\left(t_{m1(2)}\right)=\theta_2 S_y^2-\left(\theta_2-\theta_1\right)\mathbf{s}_{yx}^{/}\mathbf{S}_x^{-1}\mathbf{s}_{yx}=S_y^2\left[\theta_2-\left(\theta_2-\theta_1\right)\mathbf{s}_{yx}^{/}\mathbf{S}_x^{-1}\mathbf{s}_{yx}/S_y^2\right].$$

Now using the relation

$$\rho_{y.\mathbf{x}}^2=\mathbf{s}_{yx}^{/}\mathbf{S}_x^{-1}\mathbf{s}_{yx}/S_y^2\ , \tag{6.3.3}$$

the mean square error of $t_{m1(2)}$ is:

$$MSE\left(t_{m1(2)}\right)=\bar{Y}^2 C_y^2\left[\theta_2\left(1-\rho_{y.\mathbf{x}}^2\right)+\theta_1\rho_{y.\mathbf{x}}^2\right]. \tag{6.3.4}$$

Comparing (6.3.4) with (4.2.11) we readily see that the mean square error of $t_{m1(2)}$ is a simple extension of the mean square error of $t_{5(2)}$ as it should be.

6.3.1.2 Regression Estimator for Partial Information Case

We have given another regression estimator for two-phase sampling using a single auxiliary variable in (4.2.12). The estimator is based upon the assumption that auxiliary information is available at first phase only, along with their population information and information for the variable of interest is available at the second phase. The estimator can be thought of as an estimator for the partial information case (PIC). A simple extension of estimator $t_{6(2)}$ in the case of multiple auxiliary variables is straightforward.

$$t_{m2(2)}=\bar{y}_2+\sum_{i=1}^{q}\alpha_i\left(\bar{X}_i-\bar{x}_{(i)1}\right)=\bar{y}_2+\alpha^{/}\left(\bar{\mathbf{X}}-\bar{\mathbf{x}}_1\right). \tag{6.3.5}$$

Using notations from section 1.4, we have:

$$t_{m2(2)}-\bar{Y}=\bar{e}_{y_2}-\alpha^{/}\bar{\mathbf{e}}_{x_1}\ .$$

The mean square error of $t_{m2(2)}$ immediately may be as:

$$MSE\left(t_{m2(2)}\right)=\theta_2 S_y^2+\theta_1\alpha^{/}\mathbf{S}_x\alpha-2\theta_1\alpha^{/}\mathbf{s}_{yx}\ . \tag{6.3.6}$$

The optimum value of α which minimizes (6.3.6) is the same as given in (6.2.4). Substituting (6.2.4) in (6.3.6), simplifying and using (6.3.3), the mean square error of $t_{m2(2)}$ is:

$$MSE\left(t_{m2(2)}\right)=\bar{Y}^2 C_y^2\left(\theta_2-\theta_1\rho_{y.\mathbf{x}}^2\right), \tag{6.3.7}$$

which is an extension of (4.2.13).

The regression estimator in two-phase sampling based upon multiple auxiliary variables for PIC may also be constructed in another way. It sometimes happens that we have population information for one set of auxiliary variables and only sample information for another set of variables. The regression estimator for this situation is:

$$t_{m3(2)} = \bar{y}_2 + \sum_{i=1}^{r} \alpha_{1i}\left(\bar{x}_{(i)1} - \bar{x}_{(i)2}\right) + \sum_{i=1}^{s} \alpha_{2i}\left(\bar{W}_i - \bar{w}_{(i)1}\right),$$

or $$t_{m3(2)} = \bar{y}_2 + \alpha_1'\left(\bar{\mathrm{x}}_1 - \bar{\mathrm{x}}_2\right) + \alpha_2'\left(\bar{\mathrm{W}} - \bar{\mathrm{w}}_1\right). \quad (6.3.8)$$

As before, the estimator $t_{m3(2)}$ may be written as:

$$t_{m3(2)} - \bar{Y} = \bar{e}_{y_2} + \alpha_1'\left(\bar{\mathrm{e}}_{x_1} - \bar{\mathrm{e}}_{x_2}\right) - \alpha_2'\bar{\mathrm{e}}_{w_1}.$$

The mean square error of $t_{m3(2)}$ may be written immediately by utilizing (1.4.4) and is:

$$MSE\left(t_{m3(2)}\right) = \theta_2 S_y^2 + \left(\theta_2 - \theta_1\right)\alpha_1'\mathrm{S}_x\alpha_1 - 2\left(\theta_2 - \theta_1\right)\alpha_1'\mathrm{s}_{yx}$$
$$+\theta_1\alpha_2'\mathrm{S}_w\alpha_2 - 2\theta_1\alpha_2'\mathrm{s}_{yw}. \quad (6.3.9)$$

Differentiating (6.3.9) with respect to unknown vectors, the optimum values of α_1 and α_2 which minimizes (6.3.9) are $\alpha_1 = \mathrm{S}_x^{-1}\mathrm{s}_{yx}$ and $\alpha_2 = \mathrm{S}_w^{-1}\mathrm{s}_{yw}$. Using these optimum values in (6.3.9) and simplifying, the mean square error of $t_{m3(2)}$ is:

$$MSE\left(t_{m3(2)}\right) = \bar{Y}^2 C_y^2\left[\theta_2\left(1 - \rho_{y.\mathrm{x}}^2\right) + \theta_1\left(\rho_{y.\mathrm{x}}^2 - \rho_{y.\mathrm{w}}^2\right)\right]. \quad (6.3.10)$$

6.3.1.3 Regression Estimator for Full Information Case

We have seen in Chapter 4 that a regression estimator in two-phase sampling is easily constructed when information about the population parameter of an auxiliary variable is available. When we have information about the population mean of multiple auxiliary variables then the regression estimator in two-phase sampling may be constructed as:

$$t_{m4(2)} = \bar{y}_2 + \sum_{i=1}^{q} \alpha_i\left(\bar{X}_i - \bar{x}_{(i)2}\right) = \bar{y}_2 + \alpha'\left(\bar{\mathrm{X}} - \bar{\mathrm{x}}_2\right). \quad (6.3.11)$$

The mean square error of $t_{m4(2)}$ may immediately be written as:

$$MSE\left(t_{m3(2)}\right)=\theta_2\bar{Y}^2C_y^2\left(1-\rho_{y.\mathrm{x}}^2\right). \tag{6.3.12}$$

The mean square error given in (6.3.12) is precisely the mean square error multiple regression estimator based upon second-phase sample only.

6.3.2 Regression–cum–Ratio Estimators

We have discussed a number of estimators for two-phase sampling which combine both ratio and regression methods of estimation. Some popular estimators where both ratio and regression methods are combined have been proposed by Mohanty (1967), Kiregyera (1980), Khare and Srivastava (1981) and many others. The idea of combining ratio and regression method of estimation in two phase sampling has been further extended by Ahmad et al. (2009) when information on several auxiliary variables was available. Ahmad et al. (2009) have proposed regression–cum–ratio estimators under various situations. We discuss those estimators in the following.

6.3.2.1 Regression-Cum-Ratio Estimator for No Information Case

The regression–cum–ratio estimator proposed by Ahmad et al. (2009) for the no information case (NIC) is:

$$t_{m4(2)}=\left[\bar{y}_2+\sum_{i=1}^{r}\alpha_i\left(\bar{x}_{(i)1}-\bar{x}_{(i)2}\right)\right]\prod_{i=r+1}^{r+s=q}\left(\frac{\bar{x}_{(i)1}}{\bar{x}_{(i)2}}\right)^{\gamma_i}. \tag{6.3.13}$$

Using $\bar{x}_{(i)1}=\bar{X}_i+\bar{e}_{x_{(i)1}}$ etc. the above estimator becomes:

$$t_{m4(2)}=\left[\bar{Y}+\bar{e}_{y_2}+\sum_{i=1}^{r}\alpha_i\left(\bar{e}_{x_{(i)1}}-\bar{e}_{x_{(i)2}}\right)\right]\prod_{i=r+1}^{r+s=q}\left(1+\frac{\bar{e}_{x_{(i)1}}-\bar{e}_{x_{(i)2}}}{\bar{X}_i}\right)^{\gamma_i}.$$

Expanding and ignoring higher order terms, we have:

$$t_{m4(2)}-\bar{Y}=\bar{e}_{y_2}+\sum_{i=1}^{r}\alpha_i\left(\bar{e}_{x_{(1)i}}-\bar{e}_{x_{(2)i}}\right)+\sum_{i=r+1}^{r+s=q}\frac{\bar{Y}}{\bar{X}_i}\gamma_i\left(\bar{e}_{x_{(1)i}}-\bar{e}_{x_{(2)i}}\right).$$

Using $\alpha_i=\left(\bar{Y}/\bar{X}_i\right)\gamma_i;\left(i=r+1,\dots,q\right)$, the above equation may be put in the following form:

$$t_{m4(2)}-\bar{Y}=\bar{e}_{y_2}+\sum_{i=1}^{q}\alpha_i\left(\bar{e}_{x_{(1)i}}-\bar{e}_{x_{(2)i}}\right)=\bar{e}_{y_2}+\alpha'\left(\bar{\mathrm{e}}_{x_1}-\bar{\mathrm{e}}_{x_2}\right).$$

The mean square error of $t_{m4(2)}$ is therefore:

$$MSE\left(t_{m4(2)}\right)=E\left[\bar{e}_{y_2}+\alpha'\left(\bar{\mathrm{e}}_{x_1}-\bar{\mathrm{e}}_{x_2}\right)\right]^2 \quad .$$

Expanding the square, applying expectation and simplifying, we have:

$$MSE\left(t_{m4(2)}\right)=\theta_2 S_y^2+\left(\theta_2-\theta_1\right)\alpha'\mathrm{S}_x\alpha-2\left(\theta_2-\theta_1\right)\alpha'\mathrm{s}_{yx}. \qquad (6.3.14)$$

The expression (6.3.14) is the same as given in (6.3.2). So the optimum value of α is same as given in (6.2.4) and the mean square error of $t_{m4(2)}$ is the same as the mean square error of $t_{m1(2)}$ given in (6.3.4). Comparing mean square error expressions for regression–cum–ratio estimators for different situations with corresponding mean square error expressions of regression estimators, we conclude that regression–cum–ratio estimators, proposed by Ahmad et al. (2009), are in equivalence class with regression estimators.

6.3.2.2 Regression-Cum-Ratio Estimator for Partial Information Case

Ahmad et al. (2009) have proposed a second regression–cum–ratio estimator when population information of some of the auxiliary variables is known, as:

$$t_{m5(2)}=\left[\bar{y}_2+\sum_{i=1}^{r}\alpha_i\left(\bar{x}_{(i)1}-\bar{x}_{(i)2}\right)\right]\prod_{i=r+1}^{r+s=q}\left(\frac{\bar{x}_{(i)1}}{\bar{x}_{(i)2}}\right)^{\gamma_i}\prod_{i=r+1}^{r+s=q}\left(\frac{\bar{X}_i}{\bar{x}_{(i)1}}\right)^{\delta_i}. \qquad (6.3.15)$$

Ahmad et al. (2009) have shown that the mean square error (6.3.15) may be derived by writing it as:

$$t_{m5(2)}=\left[\bar{Y}+\bar{e}_{y_2}+\sum_{i=1}^{r}\alpha_i\left(\bar{e}_{x_{(i)1}}-\bar{e}_{x_{(i)2}}\right)\right]\prod_{i=r+1}^{r+s=q}\left(1+\frac{\bar{e}_{x_{(i)1}}-\bar{e}_{x_{(i)2}}}{\bar{X}_i}\right)^{\gamma_i}\prod_{i=r+1}^{r+s=q}\left(1+\frac{\bar{e}_{x_{(i)1}}}{\bar{X}_i}\right)^{\delta_i}.$$

Expanding and ignoring higher order terms, we have:

$$t_{m5(2)}-\bar{Y}=\bar{e}_{y_2}+\sum_{i=1}^{r}\alpha_i\left(\bar{e}_{x_{(i)1}}-\bar{e}_{x_{(i)2}}\right)+\sum_{i=r+1}^{r+s}\gamma_i\frac{\bar{Y}}{\bar{X}_i}\left(\bar{e}_{x_{(i)1}}-\bar{e}_{x_{(i)2}}\right)-\sum_{i=r+1}^{r+s}\delta_i\frac{\bar{Y}}{\bar{X}_i}\bar{e}_{x_{(i)1}}.$$

Writing $\alpha_i = \left(\bar{Y}/\bar{X}_i\right)\gamma_i;\left(i=r+1,...,q\right)$ and $\beta_i = \left(\bar{Y}/\bar{X}_i\right)\delta_i;\left(i=r+1,...,q\right)$, in the above equation we will have:

$$t_{m5(2)} - \bar{Y} = \bar{e}_{y_2} + \sum_{i=1}^{q}\alpha_i\left(\bar{e}_{x_{(i)1}} - \bar{e}_{x_{(i)2}}\right) - \sum_{i=r+1}^{r+s}\beta_i\bar{e}_{x_{(i)1}} .$$

Defining β as an $(s\times 1)$ vector and $\bar{\mathrm{e}}_{x_{(2)1}}$ as a sub vector of $\bar{\mathrm{e}}_{x_1}$, the above equation is:

$$t_{m5(2)} - \bar{Y} = \bar{e}_{y_2} + \alpha'\left(\bar{\mathrm{e}}_{x_{(1)1}} - \bar{\mathrm{e}}_{x_{(1)2}}\right) - \beta'\bar{\mathrm{e}}_{x_{(2)1}} .$$

Squaring and applying expectation, the mean square error of $t_{m5(2)}$ is:

$$MSE\left(t_{m5(2)}\right) = \theta_2 S_y^2 + \left(\theta_2 - \theta_1\right)\alpha'\mathrm{S}_x\alpha - 2\left(\theta_2 - \theta_1\right)\alpha'\mathrm{s}_{yx}$$
$$+\theta_1\beta'\mathrm{S}_{x22}\beta - 2\theta_1\beta'\mathrm{s}_{yx_{(2)}} , \qquad (6.3.17)$$

where S_{x22} is the covariance matrix for auxiliary variables $\left(X_{r+1},...,X_q\right)$ and s_{yx_2} is the vector of covariance between Y and $\left(X_{r+1},...,X_q\right)$. Immediate comparison of (6.3.16) with (6.3.9) shows that optimum values of α and β are $\alpha = \mathrm{S}_x^{-1}\mathrm{s}_{yx}$ and $\beta = \mathrm{S}_{x22}^{-1}\mathrm{s}_{yx_2}$. Substituting these values in (6.3.17) and simplifying, the mean square error of $t_{m5(2)}$ is:

$$MSE\left(t_{m5(2)}\right) = \bar{Y}^2 C_y^2\left[\theta_2\left(1-\rho_{y.\mathrm{x}}^2\right) + \theta_1\left(\rho_{y.\mathrm{x}}^2 - \rho_{y.\mathrm{x}_{(2)}}^2\right)\right]. \qquad (6.3.18)$$

Comparing the mean square error of $t_{m5(2)}$ in (6.3.18) with the mean square error of $t_{m3(2)}$ given in (6.3.10) we may say that both estimators are in equivalence class.

6.3.2.3 Regression-Cum-Ratio Estimator for Full Information Case

The regression–cum–ratio estimator, proposed by Ahmad et al. (2009), for the full information case (FIC) is:

$$t_{m6(2)} = \left[\bar{y}_2 + \sum_{i=1}^{r}\alpha_i\left(\bar{X}_i - \bar{x}_{(i)2}\right)\right]\prod_{i=r+1}^{r+s=q}\left(\frac{\bar{X}_i}{\bar{x}_{(i)2}}\right)^{\gamma_i} \qquad (6.3.19)$$

where α_i and γ_i are constants to be determined. The mean square error of $t_{m6(2)}$ may be derived by writing it as:

$$t_{m6(2)} = \left[\bar{Y} + \bar{e}_{y_2} - \sum_{i=1}^{r}\alpha_i \bar{e}_{x_{(2)i}}\right] \prod_{i=r+1}^{r+s=q}\left(1 + \frac{\bar{e}_{x_{(2)i}}}{\bar{X}_i}\right)^{-\gamma_i}.$$

Expanding and retaining terms of first order only, we have:

$$t_{m6(2)} - \bar{Y} = \bar{e}_{y_2} - \sum_{i=1}^{r}\alpha_i \bar{e}_{x_{(i)2}} - \sum_{i=r+1}^{r+s=q}\frac{\bar{Y}}{\bar{X}_i}\gamma_i \bar{e}_{x_{(i)2}}.$$

Writing $\alpha_i = \left(\bar{Y}/\bar{X}_i\right)\gamma_i; \left(i = r+1, ..., q\right)$, the above equation can be written as:

$$t_{m6(2)} - \bar{Y} = \bar{e}_{y_2} - \sum_{i=1}^{q}\alpha_i \bar{e}_{x_{(i)2}} = \bar{e}_{y_2} - \alpha' \bar{\mathrm{e}}_{x_2}. \quad (6.3.20)$$

Squaring (6.3.19), applying expectation and using section 1.4, the mean square error of $t_{m6(2)}$ is:

$$MSE\left(t_{m6(2)}\right) = \theta_2 S_y^2 + \theta_2 \alpha' \mathrm{S}_x \alpha - 2\theta_2 \alpha' \mathrm{s}_{yx}.$$

(6.3.21)

The optimum value of unknown vector ρ_{Pb} which minimizes (6.3.21) is the same as given in (6.2.4). Substituting that value in (6.3.20), we can see that the mean square error of $t_{m6(2)}$ is the same as the mean square error of $t_{m3(2)}$ which is given in (6.3.10).

6.3.3 Regression–in–Regression Estimators

The availability of multiple auxiliary variables provides us the liberty of using different combinations of estimation methods in order to increase precision in estimation. In the previous section we have discussed some estimators which combine ratio and regression methods of estimation. The ratio method of estimation can be combined with the regression method of estimation when it is supposed that some auxiliary variables may depend on another set of auxiliary variables. Ahmad et al. (2009) have proposed the regression–in–regression estimators by considering various possibilities of availability of population information of auxiliary variables. In the following we discuss these estimators for the partial information and no information cases.

6.3.3.1 Regression-in-Regression Estimator for No Information Case

Ahmad et al. (2009) have also proposed a regression–in–regression estimator for the no information case (NIC). The proposed estimator is:

$$t_{m7(2)} = \bar{y}_2 + \sum_{i=1}^{r}\alpha_{1i}\left[\left\{\bar{x}_{(i)1} + \sum_{r+1}^{r+t}\alpha_{2i}\left(\bar{x}_{(i)1} - \bar{x}_{(i)2}\right)\right\}\right.$$
$$\left.-\left\{\bar{x}_{(i)2} + \sum_{(r+t)+1}^{q}\alpha_{3i}\left(\bar{x}_{(i)1} - \bar{x}_{(i)2}\right)\right\}\right].$$

Using the vector notations, the above estimator may be written as:

$$t_{m7(2)} = \bar{y}_2 + \alpha_1'\left(\bar{\mathrm{x}}_{(1)1} - \bar{\mathrm{x}}_{(2)2}\right) + \left(\sum_{1}^{r}\alpha_{1i}\right)\alpha_2'\left(\bar{\mathrm{x}}_{(2)1} - \bar{\mathrm{x}}_{(2)2}\right)$$
$$-\left(\sum_{1}^{r}\alpha_{1i}\right)\alpha_3'\left(\bar{\mathrm{x}}_{(3)1} - \bar{\mathrm{x}}_{(3)2}\right),$$

or $t_{m7(2)} = \bar{y}_2 + \alpha_1'\left(\bar{\mathrm{x}}_{(1)1} - \bar{\mathrm{x}}_{(2)2}\right) + \beta_1'\left(\bar{\mathrm{x}}_{(2)1} - \bar{\mathrm{x}}_{(2)2}\right) + \beta_2'\left(\bar{\mathrm{x}}_{(3)1} - \bar{\mathrm{x}}_{(3)2}\right).$

(6.3.22)

In proposing this estimator, Ahmad et al. (2009) have partitioned the sample mean vector of auxiliary variables as $\bar{\mathrm{x}}_h = \left[\bar{\mathrm{x}}_{(1)h} \quad \bar{\mathrm{x}}_{(2)h} \quad \bar{\mathrm{x}}_{(3)h}\right]'$, where sub-vectors are of appropriate order. Substituting $\bar{\mathrm{x}}_{(1)1} = \bar{\mathrm{X}}_{(1)} + \bar{\mathrm{e}}_{x_{(1)1}}$ etc. in (6.3.22) we have:

$$t_{m7(2)} - \bar{Y} = \bar{e}_{y_2} + \alpha_1'\left(\bar{\mathrm{e}}_{x_{(1)1}} - \bar{\mathrm{e}}_{x_{(1)2}}\right) + \beta_1'\left(\bar{\mathrm{e}}_{x_{(2)1}} - \bar{\mathrm{e}}_{x_{(2)2}}\right) + \beta_2'\left(\bar{\mathrm{e}}_{x_{(3)1}} - \bar{\mathrm{e}}_{x_{(3)2}}\right).$$

Squaring, applying expectation and using (1.4.5), the mean square error of $t_{m7(2)}$ is:

$$MSE\left(t_{m7(2)}\right) = \theta_2 S_y^2 + (\theta_2 - \theta_1)\alpha_1'\mathrm{S}_{11}\alpha_1 + (\theta_2 - \theta_1)\beta_1'\mathrm{S}_{22}\beta_1 + (\theta_2 - \theta_1)\beta_2'\mathrm{S}_{33}\beta_2$$
$$-2(\theta_2 - \theta_1)\alpha_1'\mathrm{S}_{12}\beta_1 - 2(\theta_2 - \theta_1)\alpha_1'\mathrm{S}_{13}\beta_2 - 2(\theta_2 - \theta_1)\beta_1'\mathrm{S}_{23}\beta_2.$$

(6.3.23)

Ahmad et al. (2009) have shown that the optimum values of α_1, β_1 and β_2 which minimize (6.3.23) are $\alpha_1 = S_{11}^{-1}s_{yx_{(1)}}$, $\beta_1 = S_{22}^{-1}s_{yx_{(2)}}$ and $\beta_2 = S_{33}^{-1}s_{yx_{(3)}}$. Using these optimum values in (6.3.23), the mean square error of $t_{m7(2)}$ becomes:

$$MSE\left(t_{m7(2)}\right) = \bar{Y}^2 C_y^2 \left[\theta_2\left(1-\rho_{y.x}^2\right) + \theta_1\rho_{y.x}^2\right], \tag{6.3.24}$$

which is the same as given in (6.3.4).

6.3.3.2 Regression–in–Regression Estimator for Partial Information Case-I

Ahmad et al. (2009) have proposed five different estimators when population information of some auxiliary variables is available. The first of these five estimators is:

$$t_{m8(2)} = \bar{y}_2 + \sum_{i=1}^{r}\alpha_{1i}\left[\left\{\bar{x}_{(i)1} + \sum_{i=r+1}^{r+s=q}\alpha_{2i}\left(\bar{X}_i - \bar{x}_{(i)1}\right)\right\} - \left\{\bar{x}_{(i)2} + \sum_{i=r+1}^{r+s=q}\alpha_{3i}\left(\bar{X}_i - \bar{x}_{(i)2}\right)\right\}\right].$$

Using the vector notations the above estimator may be written as:

$$t_{m8(2)} = \bar{y}_2 + \alpha_1'\left(\bar{x}_{(1)1} - \bar{x}_{(2)1}\right) + \sum_{i=1}^{r}\alpha_{1i}\left\{\alpha_2'\left(\bar{X} - \bar{x}_{(2)1}\right)\right\} - \sum_{i=1}^{r}\alpha_{1i}\left\{\alpha_3'\left(\bar{X} - \bar{x}_{(2)2}\right)\right\}$$

,

or

$$t_{m8(2)} = \bar{y}_2 + \alpha_1'\left(\bar{x}_{(1)1} - \bar{x}_{(2)1}\right) + \beta_1'\left(\bar{X}_{(2)} - \bar{x}_{(2)1}\right) + \beta_2'\left(\bar{X}_{(2)} - \bar{x}_{(2)2}\right), \tag{6.3.25}$$

where $\bar{x}_{(1)h}$ is an $(r\times 1)$ vector of sample means at hth phase for $(X_1,\ldots,X_r)$, $\bar{X}_{(2)}$ and $\bar{x}_{(2)h}$ are $(s\times 1)$ vectors of population and sample means for $(X_{r+1},\ldots,X_q)$ respectively at h^{th} phase, $\beta_1 = \alpha_2\left(\sum_1^r \alpha_{1i}\right)$ and $\beta_2 = -\alpha_3\left(\sum_1^r \alpha_{1i}\right)$. Substituting $\bar{x}_{(1)1} = \bar{X}_{(1)} + \bar{e}_{x_{(1)1}}$ etc., the estimator (6.3.22) may be written as:

$$t_{m8(2)}-\bar{Y}=\bar{e}_{y_2}+\alpha_1'\left(\bar{e}_{x_{(1)1}}-\bar{e}_{x_{(1)2}}\right)-\beta_1'\bar{e}_{x_{(2)1}}-\beta_2'\bar{e}_{x_{(2)2}}.$$

Squaring above equation, applying expectation and using (1.4.5), the expression for mean square of $t_{m8(2)}$ is:

$$\begin{aligned}MSE\left(t_{m8(2)}\right)&=\theta_2S_y^2+\left(\theta_2-\theta_1\right)\alpha_1'\mathrm{S}_{11}\alpha_1+\theta_1\beta_1'\mathrm{S}_{22}\beta_1+\theta_2\beta_2'\mathrm{S}_{22}\beta\\&-2\left(\theta_2-\theta_1\right)\alpha_1'\mathrm{s}_{yx_{(1)}}-2\theta_1\beta_1'\mathrm{s}_{yx(2)}-2\theta_2\beta_2'\mathrm{s}_{yx(2)}\\&+2\left(\theta_2-\theta_1\right)\alpha_1'\mathrm{S}_{12}\beta_2+2\theta_1\beta_1'\mathrm{S}_{22}\beta_2.\end{aligned}\quad(6.3.26)$$

Ahmad et al. (2009) have shown that the optimum values of α_1, β_1 and β_2 which minimize (6.3.26) are $\alpha_1=\mathrm{S}_{11}^{-1}\mathrm{s}_{yx_{(1)}}$, $\beta_1=\mathrm{S}_{11}^{-1}\mathrm{s}_{yx_{(1)}}-\mathrm{S}_{22}^{-1}\mathrm{s}_{yx_{(2)}}$ and $\beta_2=\mathrm{S}_{22}^{-1}\mathrm{s}_{yx_{(2)}}$. Using these optimum values in (6.3.26) and simplifying, the mean square error of $t_{m8(2)}$ is:

$$MSE\left(t_{m8(2)}\right)=\theta_2S_y^2-\left(\theta_2-\theta_1\right)\mathrm{s}_{yx}'\mathrm{S}^{-1}\mathrm{s}_{yx}-\theta_1\mathrm{s}_{yx_{(2)}}'\mathrm{S}_{22}^{-1}\mathrm{s}_{yx_{(2)}},$$

or

$$MSE\left(t_{m8(2)}\right)=S_y^2\left[\theta_2-\left(\theta_2-\theta_1\right)\left(\mathrm{s}_{yx}'\mathrm{S}^{-1}\mathrm{s}_{yx}/S_y^2\right)-\theta_1\left(\mathrm{s}_{yx_{(2)}}'\mathrm{S}_{22}^{-1}\mathrm{s}_{yx_{(2)}}/S_y^2\right)\right].$$

Using (6.3.3) and simplifying, the mean square error of $t_{m8(2)}$ is:

$$MSE\left(t_{m8(2)}\right)=S_y^2\left[\theta_2\left(1-\rho_{y.\mathrm{x}}^2\right)+\theta_1\left(\rho_{y.\mathrm{x}}^2-\rho_{y.\mathrm{x}_{(2)}}^2\right)\right].\quad(6.3.27)$$

Comparing (6.3.27) with (6.3.17), we see that the estimators $t_{m6(2)}$ and $t_{m8(2)}$ are in equivalence class.

6.3.3.3 Regression-in-Regression Estimator for Partial Information Case-II

Ahmad et al. (2009) have proposed four other regression–in–regression estimators for partial information case. These four estimators are:

$$t_{m9(2)}=\bar{y}_2+\sum_{i=1}^{r}\alpha_{1i}\left[\left\{\bar{x}_{(i)1}+\sum_{r+1}^{q}\alpha_{2i}\left(\bar{x}_{(i)1}-\bar{x}_{(i)2}\right)\right\}\right.$$

$$-\left\{\bar{x}_{(i)2}+\sum_{r+1}^{q}\alpha_{3i}\left(\bar{X}_i-\bar{x}_{(i)2}\right)\right\}\Bigg];$$

, (6.3.28)

$$t_{m10(2)}=\bar{y}_2+\sum_{i=1}^{r}\alpha_{1i}\Bigg[\left\{\bar{x}_{(i)1}+\sum_{r+1}^{q}\alpha_{2i}\left(\bar{X}_i-\bar{x}_{(i)2}\right)\right\}$$

$$-\left\{\bar{x}_{(i)2}+\sum_{r+1}^{q}\alpha_{3i}\left(\bar{x}_{(i)1}-\bar{x}_{(i)2}\right)\right\},\tag{6.3.29}$$

$$t_{m11(2)}=\bar{y}_2+\sum_{i=1}^{r}\alpha_{1i}\Bigg[\left\{\bar{x}_{(i)1}+\sum_{r+1}^{q}\alpha_{2i}\left(\bar{x}_{(i)1}-\bar{x}_{(i)2}\right)\right\}$$

$$-\left\{\bar{x}_{(i)2}+\sum_{r+1}^{q}\alpha_{3i}\left(\bar{X}_i-\bar{x}_{(i)1}\right)\right\}\Bigg],\tag{6.3.30}$$

$$t_{m12(2)}=\bar{y}_2+\sum_{i=1}^{r}\alpha_{1i}\Bigg[\left\{\bar{x}_{(i)1}+\sum_{r+1}^{q}\alpha_{2i}\left(\bar{X}_i-\bar{x}_{(i)1}\right)\right\}$$

$$-\left\{\bar{x}_{(i)2}+\sum_{r+1}^{q}\alpha_{3i}\left(\bar{x}_{(i)1}-\bar{x}_{(i)2}\right)\right\}\Bigg].\tag{6.3.31}$$

Ahmad et al. (2009) have shown that the four estimators given in (6.3.28) to (6.3.31) have the same mean square error as the mean square error of $t_{m8(2)}$ which is given in (6.3.27).

For regression-in-regression scheme the estimator for the full information will be reduced to the regression estimator for full information case.

6.4 Ratio-Product and Regression-cum-Ratio-Product Estimators

The ratio and regression methods of estimation have been widely combined by a number of survey statisticians. In Chapter 4 we have seen some estimators which combine ratio and product methods of estimation. Some notable estimators which combine ratio and product methods of estimation are $t_{20(2)}$ proposed by Sahoo and Sahoo (1993) and given in

(4.4.11), $t_{22(2)}$ proposed by Samiuddin and Hanif (2007) and given in (4.4.13), and many others. Ahmad et al. (2009) have extended the estimators proposed by Sahoo and Sahoo (1993) and by Samiuddin and Hanif (2007) to the case of several auxiliary variables and proposed ratio–product and regression–cum–ratio–product estimators. In this section we discuss these estimators.

6.4.1 Ratio–Product Estimators

The ratio–product method of estimation has been proved to be a very useful method when some auxiliary variables have a positive correlation with the variable of study and some have a negative correlation. Ahmad et al. (2009) have proposed the ratio–product when information on several auxiliary variables is available.

6.4.1.1 Ratio-Product Estimator for No Information Case

The ratio–product estimator, proposed by Ahmad et al. (2009), when population information of none of the auxiliary variables is known is:

$$t_{m13(2)} = \bar{y}_2 \prod_{i=1}^{q} \left(\frac{\bar{x}_{(i)1}}{\bar{x}_{(i)2}} \right)^{k_i} \prod_{i=1}^{q} \left(\frac{\bar{x}_{(i)2}}{\bar{x}_{(i)1}} \right)^{1-k_i}, \tag{6.4.1}$$

where k_i are constants to be determined so that the mean square error of (6.4.1) is minimum. Substituting $\bar{x}_{(i)1} = \bar{X}_i + \bar{e}_{x_{(i)1}}$ etc. in (6.4.1), we have:

$$t_{m13(2)} = \left(\bar{Y} + \bar{e}_{y_2} \right) \left\{ 1 - \sum_{i=1}^{q} \left(\alpha_i / \bar{X}_i \right) \left(\bar{e}_{x_{(i)1}} - \bar{e}_{x_{(i)2}} \right) \right\},$$

where $\alpha_i = 1 - 2k_i$. The above equation may be written as:

$$t_{m13(2)} - \bar{Y} = \bar{e}_{y_2} - \sum_{i=1}^{q} \alpha_i \left(\bar{Y}/\bar{X}_i \right) \left(\bar{e}_{x_{(i)1}} - \bar{e}_{x_{(i)2}} \right),$$

or $$t_{m13(2)} - \bar{Y} = \bar{e}_{y_2} + \sum_{i=1}^{q} \beta_i \left(\bar{e}_{x_{(i)1}} - \bar{e}_{x_{(i)2}} \right),$$

where $\beta_i = -\alpha_i \left(\bar{Y}/\bar{X}_i \right)$. Using vector notations, the above equation may be written as:

$$t_{m13(2)} - \bar{Y} = \bar{e}_{y_2} + \beta' \left(\bar{\mathrm{e}}_{x_1} - \bar{\mathrm{e}}_{x_2} \right).$$

Squaring the above equation and applying expectation, the mean square error of $t_{13(2)}$ is:

$$MSE\left(t_{m13(2)}\right)=\theta_2 S_y^2+\left(\theta_2-\theta_1\right)\beta^{/}\mathrm{S}_x\beta-2\left(\theta_2-\theta_1\right)\beta^{/}\mathrm{s}_{yx}. \qquad (6.4.2)$$

Comparing (6.4.2) with (6.3.2), we see that the optimum value of β is the same as given in (6.2.4). We further see that the mean square error of $t_{m13(2)}$ is the same as given in (6.3.4). We therefore conclude that the ratio–product estimator for the no information case is in equivalence class with the classical regression estimator for the no information case of two phase sampling.

6.4.1.2 Ratio-Product Estimator for Partial Information Case

Ahmad et al. (2009) have proposed another ratio–product estimator when population information of some of auxiliary variables is known. The estimator is:

$$t_{m14(2)}=\bar{y}_2\prod_{i=1}^{r}\left(\frac{\bar{x}_{(i)1}}{\bar{x}_{(i)2}}\right)^{k_{1i}}\prod_{i=r+1}^{q}\left(\frac{\bar{X}_i}{\bar{x}_{(i)1}}\right)^{k_{2i}}\prod_{i=1}^{r}\left(\frac{\bar{x}_{(i)2}}{\bar{x}_{(i)1}}\right)^{1-k_{1i}}\prod_{i=1}^{r}\left(\frac{\bar{x}_{(i)1}}{\bar{X}_i}\right)^{1-k_{2i}}. \qquad (6.4.3)$$

Writing $\bar{x}_{(i)1}=\bar{X}_i+\bar{e}_{x_{(i)1}}$ etc. in (6.4.3), expanding and retaining first order terms only, we have:

$$t_{m14(2)}=\left(\bar{Y}+\bar{e}_{y_2}\right)\left[1-\sum_{i=1}^{r}\left(\alpha_{1i}/\bar{X}_i\right)\left(\bar{e}_{x_{(i)1}}-\bar{e}_{x_{(i)2}}\right)+\sum_{r+1}^{q}\left(\alpha_{2i}/\bar{X}_i\right)\bar{e}_{x_{(i)1}}\right],$$

where $\alpha_{1i}=1-2k_{1i}$ and $\alpha_{2i}=1-2k_{2i}$. The above equation may be written as:

$$t_{m14(2)}-\bar{Y}=\bar{e}_{y_2}+\sum_{i=1}^{r}\beta_{1i}\left(\bar{e}_{x_{(i)1}}-\bar{e}_{x_{(i)2}}\right)+\sum_{r+1}^{q}\beta_{2i}\bar{e}_{x_{(i)1}},$$

where $\beta_{1i}=-\alpha_{1i}\left(\bar{Y}/\bar{X}_i\right)$ and $\beta_{2i}=\alpha_{2i}\left(\bar{Y}/\bar{X}_i\right)$. The above equation may be written in a convenient form as:

$$t_{m14(2)}-\bar{Y}=\bar{e}_{y_2}+\beta_1^{/}\left(\bar{\mathrm{e}}_{x_{(1)1}}-\bar{\mathrm{e}}_{x_{(1)2}}\right)+\beta_2^{/}\bar{\mathrm{e}}_{x_{(2)1}}.$$

Squaring, applying expectation and using (1.4.5), the mean square error of $t_{m14(2)}$ is:

$$MSE\left(t_{m14(2)}\right)=\theta_2 S_y^2+\left(\theta_2-\theta_1\right)\beta_1^{/}\mathrm{S}_{11}\beta_1-2\left(\theta_2-\theta_1\right)\beta_1^{/}\mathrm{s}_{yx_{(1)}}$$
$$+\theta_1\beta_2^{/}\mathrm{S}_{22}\beta_2+2\theta_1\beta_2^{/}\mathrm{s}_{yx_{(2)}}\,. \qquad (6.4.4)$$

The optimum values of β_1 and β_2 which minimize (6.4.4) are $\beta_1=\mathrm{S}_{11}^{-1}\mathrm{s}_{yx_{(1)}}$ and $\beta_2=-\mathrm{S}_{22}^{-1}\mathrm{s}_{yx_{(2)}}$. Using optimum values of β_1 and β_2 in (6.4.4) and simplifying, the mean square error of $t_{m14(2)}$ is:

$$MSE\left(t_{m14(2)}\right)=\theta_2 S_y^2-\left(\theta_2-\theta_1\right)\mathrm{s}_{yx_{(1)}}^{/}\mathrm{S}_{11}^{-1}\mathrm{s}_{yx_{(1)}}-\theta_1\mathrm{s}_{yx_{(2)}}^{/}\mathrm{S}_{22}^{-1}\mathrm{s}_{yx_{(2)}}\,,$$

or

$$MSE\left(t_{m14(2)}\right)=\bar{Y}^2C_y^2\left[\theta_2-\left(\theta_2-\theta_1\right)\mathrm{s}_{yx_{(1)}}^{/}\mathrm{S}_{11}^{-1}\mathrm{s}_{yx_{(1)}}\Big/S_y^2-\theta_1\,\mathrm{s}_{yx_{(2)}}^{/}\mathrm{S}_{22}^{-1}\mathrm{s}_{yx_{(2)}}\Big/S_y^2\right].$$

Now using the fact that $\rho_{y.\mathrm{x}_1}^2=\mathrm{s}_{yx_{(1)}}^{/}\mathrm{S}_{11}^{-1}\mathrm{s}_{yx_{(1)}}\Big/S_y^2$ and $\rho_{y.\mathrm{x}_2}^2=\mathrm{s}_{yx_{(2)}}^{/}\mathrm{S}_{22}^{-1}\mathrm{s}_{yx_{(2)}}\Big/S_y^2$, the mean square error of $t_{m14(2)}$ may be written as:

$$MSE\left(t_{m14(2)}\right)=\bar{Y}^2C_y^2\left[\theta_2\left(1-\rho_{y.\mathrm{x}_1}^2\right)+\theta_1\left(\rho_{y.\mathrm{x}_1}^2-\rho_{y.\mathrm{x}_2}^2\right)\right]. \qquad (6.4.5)$$

Comparison of (6.4.5) with (6.3.10) shows that estimators $t_{m3(2)}$ and $t_{m14(2)}$ are in equivalence class.

6.4.1.1 Ratio-Product Estimator for Full Information Case

The ratio–product estimator, proposed by Ahmad et al. (2009), when population information of all auxiliary variables is known is:

$$t_{m15(2)}=\bar{y}_2\prod_{i=1}^{q}\left(\frac{\bar{X}_i}{\bar{x}_{(i)2}}\right)^{k_i}\prod_{i=1}^{q}\left(\frac{\bar{x}_{(i)2}}{\bar{X}_i}\right)^{1-k_i}, \qquad (6.4.6)$$

where k_i are constants to be determined so that the mean square error of (6.4.6) is minimum. Substituting $\bar{x}_{(i)1}=\bar{X}_i+\bar{e}_{x_{(i)1}}$ etc. in (6.4.6), we have:

$$t_{m15(2)} = (\bar{Y} + \bar{e}_{y_2})\left\{1 + \sum_{1=1}^{q}(\alpha_i/\bar{X}_i)(\bar{e}_{x_{(i)2}})\right\},$$

where $\alpha_i = 1 - 2k_i$. The above equation may be written as:

$$t_{m15(2)} - \bar{Y} = \bar{e}_{y_2} + \sum_{i=1}^{q}\alpha_i(\bar{Y}/\bar{X}_i)(\bar{e}_{x_{(i)2}}),$$

or $$t_{m15(2)} - \bar{Y} = \bar{e}_{y_2} - \sum_{i=1}^{q}\beta_i(\bar{e}_{x_{(i)2}}),$$

where $\beta_i = -\alpha_i\left(\bar{Y}/\bar{X}_i\right)$. Using vector notations, the above equation may be written as:

$$t_{m15(2)} - \bar{Y} = \bar{e}_{y_2} - \beta'(\bar{e}_{x_2}).$$

Squaring the above equation and applying expectation, the mean square error of $t_{m15(2)}$ is:

$$MSE(t_{m15(2)}) = \theta_2 S_y^2 + \theta_2\beta' S_x \beta - 2\theta_2\beta' s_{yx}.$$

or

$$MSE(t_{m15(2)}) = \theta_2 \bar{Y}^2 C_y^2(1 - \rho_{y.x}^2) \tag{6.4.7}$$

We further see that the mean square error of $t_{m15(2)}$ is the same as given in (6.3.12). We, therefore conclude that the ratio–product estimator for the full information case is in equivalence class with the classical regression estimator for the full information case of two phase sampling.

6.4.2 Regressions-Cum-Ratio-Product Estimators:

We have seen in section 6.3.2 how the regression method of estimation may be combined with the ratio method of estimation for developing new estimation strategies in two phase sampling. We have also seen in Chapter 4 how the regression method of estimation may be combined with the ratio and product methods of estimation to produce even more efficient estimators; the estimator $t_{56(2)}$ given in (4.10.1) is a classic example of this sort of mixing. Ahmad et al. (2010) have extended the idea presented in section 4.10 to propose estimators for two phase sampling which combine the regression method of estimation with the ratio and product methods. In this section we discuss two regression–cum–ratio–product estimators, proposed by Ahmad et al. (2010), which are based upon several auxiliary variables.

6.4.2.1 Regression-Cum-Ratio Product Estimator for Partial Information Case-I

The first regression–cum–ratio–product estimator proposed by Ahmad et al. (2010) for the partial information case is:

$$t_{m16(2)} = \left[\bar{y}_2 + \sum_{i=1}^{r} \alpha_i \left(\bar{X} - \bar{x}_{(i)1}\right)\right]\left[\prod_{r+1}^{q}\left(\frac{\bar{x}_{(i)1}}{\bar{x}_{(i)2}}\right)^{k_i} \prod_{r+1}^{q}\left(\frac{\bar{x}_{(i)2}}{\bar{x}_{(i)1}}\right)^{1-k_i}\right]. \quad (6.4.8)$$

Using the relations $\bar{x}_{(i)1} = \bar{X}_i + \bar{e}_{x_{(i)1}}$ etc., the above estimators may be written as:

$$t_{m16(2)} = \bar{Y} + \bar{e}_{y_2} - \sum_{i=1}^{r} \alpha_i \bar{e}_{x_{(i)1}} + \sum_{r+1}^{q} \beta_i \left(\bar{e}_{x_{(i)1}} - \bar{e}_{x_{(i)2}}\right),$$

where $\beta_1 = -\left(1 - 2k_i\right)\bar{Y}/\bar{X}_i$. We may write the above equation as:

$$t_{m16(2)} - \bar{Y} = \bar{e}_{y_2} - \alpha' \bar{\mathrm{e}}_{x_{(1)1}} + \beta' \left(\bar{\mathrm{e}}_{x_{(2)1}} - \bar{\mathrm{e}}_{x_{(2)2}}\right).$$

Squaring, applying expectation and using (1.4.5), the mean square error of $t_{m16(2)}$ may be written as:

$$MSE\left(t_{m16(2)}\right) = \theta_2 S_y^2 + \theta_1 \alpha' \mathrm{S}_{11} \alpha - 2\theta_1 \alpha' \mathrm{s}_{yx_{(1)}}$$
$$+ \left(\theta_2 - \theta_1\right) \beta' \mathrm{S}_{22} \beta - 2\left(\theta_2 - \theta_1\right)_1 \beta' \mathrm{s}_{yx_{(2)}}, \quad (6.4.9)$$

Ahmad et al. (2010) have shown that the optimum values of α and β which minimize (6.4.9) are $\alpha = \mathrm{S}_{11}^{-1}\mathrm{s}_{yx_{(1)}}$ and $\beta = \mathrm{S}_{22}^{-1}\mathrm{s}_{yx_{(2)}}$. Ahmad et al. (2010) have further shown that the minimum mean square of $t_{m16(2)}$ is:

$$MSE\left(t_{m16(2)}\right) = \theta_2 S_y^2 - \theta_1 \mathrm{s}'_{yx_{(1)}} \mathrm{S}_{11}^{-1} \mathrm{s}_{yx_{(1)}} - \left(\theta_2 - \theta_1\right) \mathrm{s}'_{yx_{(2)}} \mathrm{S}_{22}^{-1} \mathrm{s}_{yx_{(2)}},$$

or

$$MSE\left(t_{m16(2)}\right) = \bar{Y}^2 C_y^2 \left[\theta_2 - \left(\theta_2 - \theta_1\right) \mathrm{s}'_{yx_{(2)}} \mathrm{S}_{22}^{-1} \mathrm{s}_{yx_{(2)}} / S_y^2 - \theta_1 \mathrm{s}'_{yx_{(1)}} \mathrm{S}_{11}^{-1} \mathrm{s}_{yx_{(1)}} / S_y^2\right],$$

or

$$MSE\left(t_{m16(2)}\right)=\bar{Y}^2C_y^2\left[\theta_2\left(1-\rho_{y.x_2}^2\right)+\theta\left(\rho_{y.x_2}^2-\rho_{y.x_1}^2\right)\right]. \qquad (6.4.10)$$

Since partitioning of variables is arbitrary, through comparison of (6.4.10) with (6.4.5) we may conclude that estimators $t_{m13(2)}$, $t_{m14(2)}$ and $t_{m16(2)}$ are in equivalence class.

6.4.2.2 Regression–Cum-Ratio Product Estimator for Partial Information Case-II

The second estimator constructed by Ahmad et al. (2010) is for when population information of a set of auxiliary variables is available. The proposed estimator is:

$$t_{m17(2)}=\left[\bar{y}_2+\sum_{i=1}^{r}\alpha_i\left(\bar{x}_{(i)1}-\bar{x}_{(i)2}\right)\right]\left[\prod_{r+1}^{q}\left(\frac{\bar{X}_i}{\bar{x}_{(i)1}}\right)^{k_i}\prod_{r+1}^{q}\left(\frac{\bar{x}_{(i)1}}{\bar{X}_i}\right)^{1-k_i}\right]. \qquad (6.4.9)$$

The mean square error of $t_{m17(2)}$ may be derived by using $\bar{x}_{(i)1}=\bar{X}_i+\bar{e}_{x_{(i)1}}$ etc. in (6.4.9) and writing it as:

$$t_{m17(2)}=\bar{Y}+\bar{e}_{y_2}+\sum_{i=1}^{r}\alpha_i\left(\bar{e}_{x_{(i)1}}-\bar{e}_{x_{(i)2}}\right)-\sum_{r+1}^{q}\beta_i\bar{e}_{x_{(i)1}},$$

where $\beta_i=-\left(1-2k_i\right)\bar{Y}/\bar{X}_i$. We may write the above equation as:

$$t_{m17(2)}-\bar{Y}=\bar{e}_{y_2}+\alpha'\left(\bar{\mathrm{e}}_{x_{(1)1}}-\bar{\mathrm{e}}_{x_{(1)2}}\right)-\beta'\bar{\mathrm{e}}_{x_{(2)1}}.$$

Squaring the above equation and applying expectation, the mean square error of $t_{m17(2)}$ is:

$$MSE\left(t_{m17(2)}\right)=\theta_2S_y^2+\left(\theta_2-\theta_1\right)\alpha'\mathrm{S}_{11}\alpha-2\left(\theta_2-\theta_1\right)\alpha'\mathrm{s}_{yx_{(1)}}$$
$$+\theta_1\beta'\mathrm{S}_{22}\beta-2\theta_1\beta'\mathrm{s}_{yx_{(2)}}, \qquad (6.4.10)$$

which is the same as (6.4.4). The optimum values of unknowns are, therefore, the same as in the case of $t_{m17(2)}$ and hence the mean square error of $t_{m14(2)}$ is the same as the mean square error of $t_{m17(2)}$ and is given in (6.4.5).

 For the Regression-Cum-Ratio Product Estimators scheme, one does not not have to develop these type of estimators for Full and No information cases because they are ultimately reduced to regression estimators using multi-auxiliary variables for full and no information cases respectively at the very next step, and finally the expression for mean square error will be the same.

CHAPTER SEVEN

MULTIVARIATE ESTIMATORS IN SINGLE AND TWO-PHASE SAMPLING

7.1 Introduction

We have discussed various estimators for single and two-phase sampling using information of one, two and several auxiliary variables in the previous chapters. The ideas presented in those chapters may be extended to cases when we want to estimate several dependent variables simultaneously. This idea can be readily used in survey sampling to estimate several variables simultaneously. In this Chapter we present multivariate extensions of some popular estimators for single and two-phase sampling. The multivariate estimators will be denoted by β_1 etc., where h represent phase of sampling.

7.2 Multivariate Estimators for Single Phase Sampling

Suppose random sample of size n is drawn from a finite population of size N and information of several variables is collected in data matrix β_2. The multivariate mean per unit estimator is defined as:

$$t_{1(1)} = n^{-1}Y1 = \overline{y} = \begin{bmatrix} \overline{y}_1 & \overline{y}_2 & \cdots & \overline{y}_p \end{bmatrix}^{/}, \tag{7.2.1}$$

where $\overline{y}_j$ is sample mean of the jth variable and **1** is a p vector of 1's. Making the transformation $\overline{\mathrm{y}} = \overline{\mathrm{Y}} + \overline{\mathrm{e}}_{\mathrm{y}_1}$ we have $\mathrm{t}_{1(1)} - \overline{\mathrm{Y}} = \overline{\mathrm{e}}_{\mathrm{y}_1}$. The matrix of mean square errors of (7.2.1) is:

$$MSE\left(t_{1(1)}\right) = E\left[\left(t_{1(1)} - \overline{Y}\right)\left(t_{1(1)} - \overline{Y}\right)^{/}\right] = E\left(\overline{e}_{y_1}\overline{e}_{y_1}^{/}\right).$$

Using (1.4.6), the mean square error of $t_{1(1)}$ is:

$$MSE\left(t_{1(1)}\right) = \theta_1 S_y . \tag{7.2.2}$$

The mean square error given in (7.2.2) reduces to the mean square error of univariate mean per unit estimator for $p = 1$.

The multivariate regression estimator for single-phase sampling may be readily defined parallel to lines of multivariate regression model. The estimator for several variables of interest is:

$$t_{2(1)} = \bar{y} + B\left(\bar{X} - \bar{x}\right), \tag{7.2.3}$$

where $\bar{\mathrm{y}}$ is a $(p \times 1)$ vector of sample means of variables of interest, **B** is a $(p \times q)$ matrix of unknown coefficients, $\bar{X}$ is a $(q \times 1)$ vector of population means of auxiliary variables and $\bar{\mathrm{x}}$ is a $(q \times 1)$ vector of sample means of auxiliary variables. Using $\bar{y} = \bar{Y} + \bar{e}_{y_1}$ etc. in (7.2.3) we have:

$$t_{2(1)} - \bar{Y} = \bar{e}_{y_1} - B\bar{e}_{x_1} .$$

The mean square error of $t_{2(1)}$ is:

$$MSE\left(\mathbf{t}_{2(1)}\right) = E\left[\left(\mathbf{t}_{2(1)} - \bar{\mathbf{Y}}\right)\left(\mathbf{t}_{2(1)} - \bar{\mathbf{Y}}\right)'\right] = E\left[\left(\bar{\mathbf{e}}_{\mathbf{y}_1} - \mathbf{B}\bar{\mathbf{e}}_{\mathbf{x}_1}\right)\left(\bar{\mathbf{e}}_{\mathbf{y}_1} - \mathbf{B}\bar{\mathbf{e}}_{\mathbf{x}_1}\right)'\right]$$

Multiplying and using (1.4.6), the mean square error of ρ_{Pb} is:

$$MSE\left(\mathrm{t}_{2(1)}\right) = \theta_1 \mathrm{S_y} - \theta_1 \mathrm{BS_{xy}} - \theta_1 \mathrm{S_{yx}B'} + \theta_1 \mathrm{BS_x B'} . \tag{7.2.4}$$

Partially differentiating (7.2.4) and equating to zero, the optimum value of **B** which minimizes mean square error of $t_{2(1)}$ is $B = S_{yx} S_x^{-1}$ which is matrix of regression coefficients of **y** on **x**. Using optimum value of **B** in (7.2.4) and simplifying, the mean square error of $\mathrm{t}_{2(1)}$ is:

$$MSE\left(t_{2(1)}\right) \approx \theta_1 \left(S_y - S_{yx} S_x^{-1} S_{xy}\right) . \tag{7.2.5}$$

It may be immediately seen that the mean square error (7.2.5) is an extension of the mean square error of the classical regression estimator given in (6.2.5). This can be easily verified by using the fact that for a single variable of interest $\bar{y}$ will have just a single entry. Also for this case various matrices given in (7.2.5) will be:

$$b = \frac{-\bar{Y}C_y}{PC_\tau}\rho_{pb} ,$$

and

$$\mathbf{S_x} = \begin{bmatrix} s_{x_1}^2 & s_{x_1x_2} & \cdots & s_{x_1x_q} \\ s_{x_2x_1} & s_{x_2}^2 & \cdots & s_{x_2x_q} \\ \vdots & \vdots & \ddots & \vdots \\ s_{x_qx_1} & s_{x_qx_2} & \cdots & s_{x_q}^2 \end{bmatrix}.$$

Using these matrices in (7.2.5) and simplifying we have (6.2.5). The mean square error (7.2.5) may also be used to derive the covariance between two estimates. For illustration, suppose $p = 2$ and $q = 2$. In this case various matrices in (7.2.5) are:

$$\left.\begin{aligned} \mathbf{S_y} &= \begin{bmatrix} S_{y_1}^2 & \rho_{y_1y_2}S_{y_1}S_{y_2} \\ & S_{y_2}^2 \end{bmatrix}; \mathbf{S_x} = \begin{bmatrix} S_{x_1}^2 & \rho_{x_1x_2}S_{x_1}S_{x_2} \\ & S_{x_2}^2 \end{bmatrix} \\ \mathbf{S_{yx}} &= \begin{bmatrix} \rho_{y_1x_1}S_{y_1}S_{x_1} & \rho_{y_1x_2}S_{y_1}S_{x_2} \\ \rho_{y_2x_1}S_{y_2}S_{x_1} & \rho_{y_2x_2}S_{y_2}S_{x_2} \end{bmatrix} \end{aligned}\right\}. \tag{7.2.6}$$

Using (7.2.6) in (7.2.5) and simplifying, the covariance between two estimators is:

$$\begin{aligned} Cov(t_1,t_2) = S_{y_1}S_{y_2}\Big[\rho_{y_1y_2} + \left(1-\rho_{x_1x_2}^2\right)^{-1}\big\{\rho_{y_1x_1}\rho_{y_2x_1} + \rho_{y_1x_2}\rho_{y_2x_2} \\ -\rho_{x_1x_2}\left(\rho_{y_1x_2}\rho_{y_2x_1} + \rho_{y_1x_1}\rho_{y_2x_2}\right)\big\}\Big]. \end{aligned} \tag{7.2.7}$$

Expression (7.2.5) has wide applicability and may be used to obtain the mean square error expression for any number of estimates.

7.3 Multivariate Regression Estimators for Two-Phase Sampling

We have discussed the regression estimator for two-phase sampling with single and two auxiliary variables in Chapter 4. The regression estimator for two-phase sampling with multiple auxiliary variables was discussed in Chapter 6. We know that in the case of multiple auxiliary variables, in two-phase sampling, the regression estimator may be constructed using different possibilities which are termed as the No Information Case, Partial Information Case, and Full Information Case. We now discuss a multivariate regression estimator for two phase sampling as an extension of estimators discussed in Chapter 6.

7.3.1 Multivariate regression Estimator for No Information Case

The multivariate regression estimator for two-phase sampling when only sample information is available has been proposed by Ahmad et al. (2010) as:

$$t_{1(2)} = \overline{y}_2 + B(\overline{x}_1 - \overline{x}_2), \tag{7.3.1}$$

where $\overline{y}_2$ is $(p \times 1)$ vector of the means of the variables of interest for second phase sample, **B** is $(q \times p)$ matrix of unknown coefficients, $\overline{x}_1$ and $\overline{x}_2$ are $(q \times 1)$ sample mean vectors of auxiliary variables for first phase and second phase sample respectively. Using $\overline{y} = \overline{Y} + \overline{e}_{y_2}$ etc. in (7.3.1) we have:

$$t_{1(2)} - \overline{Y} = \overline{e}_{y_2} + B\left(\overline{e}_{x_1} - \overline{e}_{x_2}\right). \tag{7.3.2}$$

Squaring (7.3.2), applying expectation and using (1.4.6), the mean square error of $t_{1(2)}$ is:

$$MSE\left(t_{1(2)}\right) = \theta_2 S_y - (\theta_2 - \theta_1) BS_{xy} - (\theta_2 - \theta_1) S_{yx} B' + (\theta_2 - \theta_1) BS_x B'. \tag{7.3.3}$$

Ahmad et al. (2010) have shown that the optimum value of **B** which minimizes (7.3.3) is $B = S_{yx}S_x^{-1}$ which contains regression coefficients of Y's on X's. Substituting the optimum value of **B** in (7.3.3) and simplifying, the mean square error of $\overline{Y}$ is:

$$MSE\left(t_{1(2)}\right) \approx \theta_2 \left(S_y - S_{yx} S_x^{-1} S_{xy}\right) + \theta_1 S_{yx} S_x^{-1} S_{xy}. \tag{7.3.4}$$

The mean square error expression given in (7.3.4) may be used to compute mean square error for any number of study variables. Ahmad et al. (2010) have shown that the mean square error for the classical regression estimator of two phase sampling with one, two and several auxiliary variables are special cases of (7.3.4). We illustrate this with some examples.

If we use $p = 1$ and $q = 2$ in (7.3.1) then we have a two phase sampling regression estimator for the no information case. The mean square error expression may be obtained by using the following matrices in (7.3.4):

$$S_y = \left[S_y^2\right]; S_{yx} = \left[S_{yx_1} \quad S_{yx_2} \quad \cdots \quad S_{yx_q}\right],$$

and

$$S_x = \begin{bmatrix} S_{x_1}^2 & S_{x_1x_2} & \cdots & S_{x_1x_q} \\ S_{x_2x_1} & S_{x_2}^2 & \cdots & S_{x_2x_q} \\ \vdots & \vdots & \ddots & \vdots \\ S_{x_qx_1} & S_{x_qx_2} & \cdots & S_{x_q}^2 \end{bmatrix}.$$

Using these matrices in (7.3.4) and simplifying, the mean square error of the regression estimator with two auxiliary variables is readily found as:

$$MSE\left(t_{1(2)}\right) \approx S_y^2\left[\theta_2\left(1-\rho_{y.x_1x_2}^2\right)+\theta_1\rho_{y.x_1x_2}^2\right]. \tag{7.3.5}$$

Again using $p = 2$ and $q = 1$ in (7.3.1) we have a pair of regression estimators. The mean square error matrix of the pair may be obtained by using the following matrices:

$$\mathrm{S_y} = \begin{bmatrix} S_{y_1}^2 & \rho_{y_1y_2}S_{y_1}S_{y_2} \\ & S_{y_2}^2 \end{bmatrix}; \mathrm{S_{yx}} = \begin{bmatrix} \rho_{y_1x}S_{y_1}S_x \\ \rho_{y_2x}S_{y_2}S_x \end{bmatrix}; \mathrm{S_x} = \left[S_x^2\right]. \tag{7.3.6}$$

Using (7.3.6) in (7.3.4) and simplifying, the mean square error matrix of pair of estimates is:

$$MSE\left(\mathrm{t}_{1(2)}\right) \approx \theta_2\begin{bmatrix} S_{y_1}^2\left(1-\rho_{y_1x}^2\right) & S_{y_1}S_{y_2}\left(\rho_{y_1y_2}-\rho_{y_1x}\rho_{y_2x}\right) \\ & S_{y_2}^2\left(1-\rho_{y_2x}^2\right) \end{bmatrix}$$

$$\theta_1\begin{bmatrix} S_{y_1}^2\rho_{y_1x}^2 & S_{y_1}S_{y_2}\rho_{y_1x}\rho_{y_2x} \\ & S_{y_2}^2\rho_{y_2x}^2 \end{bmatrix}. \tag{7.3.7}$$

Expression (7.3.7) immediately shows that the covariance between two estimates is:

$$Cov\left(t_1,t_2\right) = S_{y_1}S_{y_2}\left[\theta_2\left(\rho_{y_1y_2}-\rho_{y_1x}\rho_{y_2x}\right)+\theta_1\rho_{y_1x}\rho_{y_2x}\right]. \tag{7.3.8}$$

The expression (7.3.4) may be used to obtain the mean square error matrix for any number of study variables.

7.3.2 Multivariate Regression Estimator for Partial Information Case

Ahmad et al. (2010) have proposed various multivariate regression estimators for the partial information case also. Before we give the estimator for the partial information case, let us assume that the vector of auxiliary variables has been partitioned into two sub-vectors. Corresponding to the partitioning of the vector of auxiliary variables, the vectors of e's are also partitioned, which gives rise to following partitioning of covariance matrices:

$$S_{yx}=\begin{bmatrix} S_{yx_{(1)}} & S_{yx_{(2)}} \end{bmatrix}; S_x=\begin{bmatrix} S_{x_{(11)}} & S_{x_{(12)}} \\ S_{x_{(21)}} & S_{x_{(22)}} \end{bmatrix}=\begin{bmatrix} S_{11} & S_{12} \\ S_{21} & S_{22} \end{bmatrix}. \tag{7.3.9}$$

We now give the multivariate regression estimator for the partial information case, proposed by Ahmad et al. (2010), which is:

$$t_{2(2)}=\bar{y}_2+B_1\left(\bar{x}_{(1)1}-\bar{x}_{(1)2}\right)+B_2\left(\bar{X}_{(2)}-\bar{x}_{(2)1}\right)+B_3\left(\bar{X}_{(2)}-\bar{x}_{(2)2}\right), \tag{7.3.10}$$

where matrices are of appropriate order. Using $g_{\omega}\left(\bar{y},v\right)=\bar{y}\left[k\,v^{\alpha}+\left(1-k\right)e^{\alpha\left(v-1\right)}\right]$ etc., the estimator (7.3.10) may be written as:

$$t_{2(2)}-\bar{Y}=\bar{e}_{y_2}+B_1\left(\bar{e}_{x_{(1)1}}-\bar{e}_{x_{(1)2}}\right)-B_2\bar{e}_{x_{(2)1}}-B_3\bar{e}_{x_{(2)2}}. \tag{7.3.11}$$

The mean square error matrix of (7.3.10) may be obtained by squaring (7.3.11), applying expectation and using (1.4.6) and (7.3.9). Ahmad et al. (2010) have shown that the mean square error matrix of (7.3.10) may be written as:

$$\begin{aligned} MSE\left(t_{2(2)}\right) &\approx \theta_2 S_y-\left(\theta_2-\theta_1\right)B_1S_{xy}-\theta_1B_2S_{x_{(2)}y}-\theta_2B_3S_{x_{(2)}y} \\ &-\left(\theta_2-\theta_1\right)S_{yx(1)}B_1^{/}+\left(\theta_2-\theta_1\right)B_1S_{11}B_1^{/}+\left(\theta_2-\theta_1\right)B_3S_{21}B_1^{/} \\ &-\theta_1S_{yx_{(2)}}B_2^{/}+\theta_1B_2S_{22}B_2^{/}+\theta_1B_3S_{22}B_2^{/}-\theta_2S_{yx_{(2)}}B_3^{/} \\ &+\left(\theta_2-\theta_1\right)B_1S_{12}B_3^{/}+\theta_1B_2S_{22}B_3^{/}+\theta_2B_3S_{22}B_3^{/}. \end{aligned} \tag{7.3.12}$$

Partially differentiating (7.3.12) *w.r.t.* unknown coefficients and equating to zero, Ahmad et al. (2010) have obtained the following optimum values of $\mathbf{B}_1$, $\mathbf{B}_2$ and $\mathbf{B}_3$ which minimize (7.3.12):

$$\left.\begin{aligned} B_1 &= S_{yx_{(1)}} S_{11}^{-1} \\ B_2 &= S_{yx_{(1)}} S_{11}^{-1} - S_{yx_{(2)}} S_{22}^{-1} \\ B_3 &= S_{yx_{(2)}} S_{22}^{-1} \end{aligned}\right\}. \tag{7.3.13}$$

Using (7.3.13) in (7.3.12) and simplifying, Ahmad et al. (2010) have obtained the following mean square error of $t_{2(2)}$:

$$MSE\left(t_{2(2)}\right) \approx \left[\theta_2\left(S_y - S_{yx}S_x^{-1}S_{xy}\right) + \theta_1\left(S_{yx}S_x^{-1}S_{xy} - S_{yx_{(2)}}S_{22}^{-1}S_{x_{(2)}y}\right)\right]. \tag{7.3.14}$$

The expression (7.3.16) is a general expression to obtain the mean square error of any component of estimator vector $t_{2(2)}$. As an illustration suppose $p = 1$, then estimator (7.3.10) reduces to estimator $t_{m9(2)}$ given in (6.3.24). The mean square of (6.3.24) given in (6.3.26) may be obtained from (7.3.14) by using the following matrices in (7.3.1 4):

$$S_y = \left[s_y^2\right]; S_{yx} = \left[s_{yx_{(1)}} \quad s_{yx_{(2)}}\right]; S_x = \begin{bmatrix} S_{11} & S_{12} \\ S_{21} & S_{22} \end{bmatrix},$$

where $\mathbf{s}_{y\mathbf{x}}$ is now a $(1\times p)$ vector. Now, various entries of (7.3.14) are:

$$\left.\begin{aligned} S_{yx}S_x^{-1}S_{xy} &= s_{yx}S_x^{-1}s_{yx}^{/} = S_y^2\rho_{y.x}^2 \\ S_{yx_{(2)}}S_{22}^{-1}S_{x_{(2)}y} &= s_{yx_{(2)}}S_{22}^{-1}s_{yx_{(2)}}^{/} = S_y^2\rho_{y.x_{(2)}}^2 \end{aligned}\right\}. \tag{7.3.15}$$

Putting (7.3.15) in (7.3.14) and simplifying we have:

$$MSE\left(\mathrm{t}_{2(2)}\right) = S_y^2\left[\theta_2\left(1-\rho_{y.\mathrm{x}}^2\right) + \theta_1\left(\rho_{y.\mathrm{x}}^2 - \rho_{y.\mathrm{x}_{(2)}}^2\right)\right], \tag{6.3.26}$$

which is (6.3.23). The expression (7.3.16) may be used to obtain the mean square error for any number of study variables.

7.3.3 Multivariate Regression Estimator for Full Information Case

Ahmad et al. (2010) have also proposed the multivariate regression estimator in two-phase sampling when population information of all auxiliary variables is known; the Full Information Case. The estimator vector for population mean in this case is:

$$t_{3(2)} = \bar{y}_2 + B\left(\bar{X} - \bar{x}_2\right), \tag{7.3.16}$$

where notations have their usual meanings. Ahmad et al. (2010) have shown that the mean square error matrix for (7.3.16) may be easily obtained as it is obtained in the case of the multivariate regression estimator in single phase sampling. The mean square error matrix of (7.3.16) obtained by Ahmad et al. (2010) is:

$$MSE\left(t_{3(2)}\right) = \theta_2\left(S_y - S_{yx}S_x^{-1}S_{xy}\right). \tag{7.3.17}$$

The expression (7.3.17) is again an extension of the mean square error of the univariate regression estimator for the Full Information Case given in (6.3.12). Comparison of (7.3.4) with (7.3.17) also shows that the estimator for the full information case is more precise as compared with the estimator for the no information case.

7.4 Multivariate Ratio Estimators for Two-Phase Sampling

The ratio estimator for single and two phase sampling has been studied by various authors using information of single and multiple auxiliary variables. Availability of information of multiple auxiliary variables has also given rise to various situations for the use of ratio estimators which are discussed by Samiuddin and Hanif (2007). The ratio estimator for several study variables has been analogously proposed by Hanif et al. (2009) when information on multiple auxiliary variables is available. Hanif et al. (2009) have proposed a ratio estimator for no information, partial and full information cases.

7.4.1 Multivariate Ratio Estimator for No Information case

The multivariate ratio estimator in two phase sampling for the no information case is given as:

$$t_{4(2)}=\left[\bar{y}_{(1)2}\prod_{i=1}^{q}\left(\frac{\bar{x}_{(i)1}}{\bar{x}_{(i)2}}\right)^{\alpha_{i1}}\quad \bar{y}_{(2)2}\prod_{i=1}^{q}\left(\frac{\bar{x}_{(i)1}}{\bar{x}_{(i)2}}\right)^{\alpha_{i1}}\quad \cdots\quad \bar{y}_{(p)2}\prod_{i=1}^{q}\left(\frac{\bar{x}_{(i)1}}{\bar{x}_{(i)2}}\right)^{\alpha_{i1}}\right]^{/}. \tag{7.4.1}$$

Using $\bar{y}_{(i)2}=\bar{Y}_i+\bar{e}_{y_{(i)2}}$ etc., expanding and retaining linear terms only, the estimator (7.4.1) may be written as:

$$t_{4(2)}=\begin{bmatrix}\left(\bar{Y}_1+\bar{e}_{y_{(2)2}}\right)\left\{1-\sum_{j=1}^{q}\frac{\alpha_{1j}}{\bar{X}_j}\left(\bar{e}_{x_{(j)1}}-\bar{e}_{x_{(j)2}}\right)\right\}\\ \left(\bar{Y}_2+\bar{e}_{y_{(2)2}}\right)\left\{1-\sum_{j=1}^{q}\frac{\alpha_{2j}}{\bar{X}_j}\left(\bar{e}_{x_{(j)1}}-\bar{e}_{x_{(j)2}}\right)\right\}\\ \vdots\\ \vdots\\ \left(\bar{Y}_1+\bar{e}_{y_{(p)2}}\right)\left\{1-\sum_{j=1}^{q}\frac{\alpha_{pj}}{\bar{X}_j}\left(\bar{e}_{x_{(j)1}}-\bar{e}_{x_{(j)2}}\right)\right\}\end{bmatrix},$$

$$\text{or}\quad t_{4(2)}=\begin{bmatrix}\left(\bar{Y}_1+\bar{e}_{y_{(1)2}}\right)\\ \left(\bar{Y}_2+\bar{e}_{y_{(2)2}}\right)\\ \vdots\\ \vdots\\ \left(\bar{Y}_1+\bar{e}_{y_{(p)2}}\right)\end{bmatrix}-\begin{bmatrix}\bar{Y}_1\sum_{j=1}^{q}\frac{\alpha_{1j}}{\bar{X}_j}\left(\bar{e}_{x_{(j)1}}-\bar{e}_{x_{(j)2}}\right)\\ \bar{Y}_2\sum_{j=1}^{q}\frac{\alpha_{2j}}{\bar{X}_j}\left(\bar{e}_{x_{(j)1}}-\bar{e}_{x_{(j)2}}\right)\\ \vdots\\ \vdots\\ \bar{Y}_p\sum_{j=1}^{q}\frac{\alpha_{pj}}{\bar{X}_j}\left(\bar{e}_{x_{(j)1}}-\bar{e}_{x_{(j)2}}\right)\end{bmatrix},$$

Using vector notations, the above estimator may be written as:

$$t_{4(2)}-\bar{Y}=\bar{e}_y+A\left(\bar{e}_{x_1}-\bar{e}_{x_2}\right), \tag{7.4.2}$$

where **A** is a $(p \times q)$ matrix with entries $A = \left[-\left(\bar{Y}_i/\bar{X}_j\right)\alpha_{ij}\right]$. Squaring (7.4.2) and applying expectation, the mean square error of $t_{4(2)}$ is:

$$MSE\left(t_{4(2)}\right) \approx \theta_2 S_y - (\theta_2 - \theta_1) A S_{xy} - (\theta_2 - \theta_1) S_{yx} A' + (\theta_2 - \theta_1) A S_x A'. \tag{7.4.3}$$

Hanif et al. (2009) have shown that the optimum value of **A** which minimizes (7.4.3) is $A = S_{yx} S_x^{-1}$. Using this optimum value of **A** in (7.4.3) and simplifying, the mean square error matrix of $t_{4(2)}$ is:

$$MSE\left(t_{4(2)}\right) \approx \theta_2 \left(S_y - S_{yx} S_x^{-1} S_{xy}\right) + \theta_1 S_{yx} S_x^{-1} S_{xy}. \tag{7.4.4}$$

Comparing (7.4.4) with (7.3.4) we conclude that the estimator $t_{1(2)}$ and $t_{4(2)}$ are in the same equivalence class, as both have the same mean square error matrix.

7.4.2 Multivariate Ratio Estimator for Partial Information case

Earlier we gave the regression estimator when population information of some of the auxiliary variables is available. This situation has been termed the partial information case by Samiuddin and Hanif (2007). Ahmad et al. (2010) have proposed regression–cum–ratio–product estimators for this type of situation and we have discussed these estimators in section 6.4.2. Ahmad et al. (2010) have further extended the ratio estimator for estimation of several variables simultaneously. The proposed estimator has been termed as the multivariate ratio estimator for the partial information case. The estimator may be seen as a collection of several ratio estimators which are based upon the same set of predictors. The proposed estimator is:

$$\mathrm{t}_{5(2)}=\begin{bmatrix}\bar{y}_{(1)2}\prod_{j=1}^{r}\left(\frac{\bar{x}_{(j)1}}{\bar{x}_{(j)2}}\right)^{\alpha_{1j}}\prod_{j=r+1}^{q}\left(\frac{\bar{X}_j}{\bar{x}_{(j)1}}\right)^{\beta_{1j}}\prod_{j=r+1}^{q}\left(\frac{\bar{X}_j}{\bar{x}_{(j)2}}\right)^{\gamma_{1j}}\\ \bar{y}_{(2)2}\prod_{j=1}^{r}\left(\frac{\bar{x}_{(j)1}}{\bar{x}_{(j)2}}\right)^{\alpha_{2j}}\prod_{j=r+1}^{q}\left(\frac{\bar{X}_j}{\bar{x}_{(j)1}}\right)^{\beta_{2j}}\prod_{j=r+1}^{q}\left(\frac{\bar{X}_j}{\bar{x}_{(j)2}}\right)^{\gamma_{2j}}\\ \vdots\\ \vdots\\ \bar{y}_{(p)2}\prod_{j=1}^{r}\left(\frac{\bar{x}_{(j)1}}{\bar{x}_{(j)2}}\right)^{\alpha_{pj}}\prod_{j=r+1}^{q}\left(\frac{\bar{X}_j}{\bar{x}_{(j)1}}\right)^{\beta_{pj}}\prod_{j=r+1}^{q}\left(\frac{\bar{X}_j}{\bar{x}_{(j)2}}\right)^{\gamma_{pj}}\end{bmatrix}. \quad (7.4.5)$$

The estimator (7.4.5) uses information at the first and second phase of sampling, alongside population information of the auxiliary variables. The mean square error matrix of (7.4.5) may be derived by writing $\bar{y}_{(i)2}=\bar{Y}_i+\bar{e}_{y_{(i)2}}$ etc., expanding and retaining linear terms only. Under these transformations, the estimator (7.4.5) may be written as:

$$\mathrm{t}_{5(2)}=\begin{bmatrix}\left(\bar{Y}_1+\bar{e}_{y_{(1)2}}\right)\left\{1+\sum_{j=1}^{r}\frac{\alpha_{1j}}{\bar{X}_j}\left(\bar{e}_{x_{(j)1}}-\bar{e}_{x_{(j)2}}\right)\right\}\left\{1-\sum_{j=r+1}^{q}\frac{\beta_{1j}}{\bar{X}_j}\bar{e}_{x_{(j)1}}\right\}\left\{1-\sum_{i=r+1}^{q}\frac{\gamma_{1j}}{\bar{X}_j}\bar{e}_{x_{(j)2}}\right\}\\ \left(\bar{Y}_1+\bar{e}_{y_{(1)2}}\right)\left\{1+\sum_{j=1}^{r}\frac{\alpha_{2j}}{\bar{X}_j}\left(\bar{e}_{x_{(j)1}}-\bar{e}_{x_{(j)2}}\right)\right\}\left\{1-\sum_{j=r+1}^{q}\frac{\beta_{2j}}{\bar{X}_j}\bar{e}_{x_{(j)1}}\right\}\left\{1-\sum_{i=r+1}^{q}\frac{\gamma_{2j}}{\bar{X}_j}\bar{e}_{x_{(j)2}}\right\}\\ \vdots\\ \vdots\\ \left(\bar{Y}_1+\bar{e}_{y_{(1)2}}\right)\left\{1+\sum_{j=1}^{r}\frac{\alpha_{pj}}{\bar{X}_j}\left(\bar{e}_{x_{(j)1}}-\bar{e}_{x_{(j)2}}\right)\right\}\left\{1-\sum_{j=r+1}^{q}\frac{\beta_{pj}}{\bar{X}_j}\bar{e}_{x_{(j)1}}\right\}\left\{1-\sum_{i=r+1}^{q}\frac{\gamma_{pj}}{\bar{X}_j}\bar{e}_{x_{(j)2}}\right\}\end{bmatrix}.$$

or

$$\mathrm{t}_{5(2)}=\begin{bmatrix}\left(\bar{Y}_1+\bar{e}_{y_{(1)2}}\right)\\ \left(\bar{Y}_1+\bar{e}_{y_{(1)2}}\right)\\ \vdots\\ \vdots\\ \left(\bar{Y}_1+\bar{e}_{y_{(1)2}}\right)\end{bmatrix}+\begin{bmatrix}\sum_{j=1}^{r}\frac{\alpha_{1j}}{\bar{X}_j}\left(\bar{e}_{x_{(j)1}}-\bar{e}_{x_{(j)2}}\right)\\ \sum_{j=1}^{r}\frac{\alpha_{2j}}{\bar{X}_j}\left(\bar{e}_{x_{(j)1}}-\bar{e}_{x_{(j)2}}\right)\\ \vdots\\ \vdots\\ \sum_{j=1}^{r}\frac{\alpha_{pj}}{\bar{X}_j}\left(\bar{e}_{x_{(j)1}}-\bar{e}_{x_{(j)2}}\right)\end{bmatrix}-\begin{bmatrix}\sum_{j=r+1}^{q}\frac{\beta_{1j}}{\bar{X}_j}\bar{e}_{x_{(j)1}}\\ \sum_{j=r+1}^{q}\frac{\beta_{2j}}{\bar{X}_j}\bar{e}_{x_{(j)1}}\\ \vdots\\ \vdots\\ \sum_{j=r+1}^{q}\frac{\beta_{pj}}{\bar{X}_j}\bar{e}_{x_{(j)1}}\end{bmatrix}-\begin{bmatrix}\sum_{i=r+1}^{q}\frac{\gamma_{1j}}{\bar{X}_j}\bar{e}_{x_{(j)2}}\\ \sum_{i=r+1}^{q}\frac{\gamma_{2j}}{\bar{X}_j}\bar{e}_{x_{(j)2}}\\ \vdots\\ \vdots\\ \sum_{i=r+1}^{q}\frac{\gamma_{pj}}{\bar{X}_j}\bar{e}_{x_{(j)2}}\end{bmatrix}.$$

Using matrix representations, the above equation may be written as:

$$\mathrm{t}_{5(2)}-\bar{\mathrm{Y}}=\bar{\mathrm{e}}_{\mathrm{y}_2}+\mathrm{A}_1\left(\bar{\mathrm{e}}_{\mathrm{x}_{(1)1}}-\bar{\mathrm{e}}_{\mathrm{x}_{(1)2}}\right)-\mathrm{A}_2\bar{\mathrm{e}}_{\mathrm{x}_{(2)1}}-\mathrm{A}_3\bar{\mathrm{e}}_{\mathrm{x}_{(2)2}}, \qquad (7.4.6)$$

where $\mathrm{A}_{1(p\times r)}=\left[\left(\bar{Y}_i/\bar{X}_j\right)\alpha_{ij}\right]$, $\mathrm{A}_{2(p\times s)}=\left[\left(\bar{Y}_i/\bar{X}_j\right)\beta_{ij}\right]$ and $\mathrm{A}_{3(p\times s)}=\left[\left(\bar{Y}_i/\bar{X}_j\right)\gamma_{ij}\right]$ are coefficient vectors. The mean square error matrix of $\mathrm{t}_{5(2)}$ is:

$$MSE\left(\mathrm{t}_{5(2)}\right)=E\left[\left(\mathrm{t}_{6(2)}-\bar{\mathrm{Y}}\right)\left(\mathrm{t}_{6(2)}-\bar{\mathrm{Y}}\right)^{/}\right].$$

Using (7.4.6) in the above equation, using (7.3.11) and simplifying, the mean square error matrix of $\mathrm{t}_{5(2)}$ is:

$$MSE\left(\mathrm{t}_{5(2)}\right) \approx \theta_2 \mathrm{S_y} - (\theta_2 - \theta_1)\mathrm{A_1 S_{xy}} - \theta_1 \mathrm{A_2 S_{x_{(2)}y}} - \theta_2 \mathrm{A_3 S_{x_{(2)}y}}$$
$$-(\theta_2 - \theta_1)\mathrm{S_{yx(1)}A_1'} + (\theta_2 - \theta_1)\mathrm{A_1 S_{11} A_1'} + (\theta_2 - \theta_1)\mathrm{A_3 S_{21} A_1'}$$
$$-\theta_1 \mathrm{S_{yx_{(2)}} A_2'} + \theta_1 \mathrm{A_2 S_{22} A_2'} + \theta_1 \mathrm{A_3 S_{22} A_2'} - \theta_2 \mathrm{S_{yx_{(2)}} A_3'}$$
$$+(\theta_2 - \theta_1)\mathrm{A_1 S_{12} A_3'} + \theta_1 \mathrm{A_2 S_{22} A_3'} + \theta_2 \mathrm{A_3 S_{22} A_3'} \quad . \qquad (7.4.7)$$

Ahmad et al. (2010) have shown that the optimum values of $\mathbf{A}_1$, $\mathbf{A}_2$ and $\mathbf{A}_3$ which minimize (7.3.14):

$$\left.\begin{aligned} \mathrm{A_1} &= \mathrm{S_{yx_{(1)}} S_{11}^{-1}} \\ \mathrm{A_2} &= \mathrm{S_{yx_{(1)}} S_{11}^{-1} - S_{yx_{(2)}} S_{22}^{-1}} \\ \mathrm{A_3} &= \mathrm{S_{yx_{(2)}} S_{22}^{-1}} \end{aligned}\right\} . \qquad (7.4.8)$$

Using (7.4.8) in (7.4.7) and simplifying, Ahmad et al. (2010) have obtained following mean square error of $\mathrm{t}_{5(2)}$:

$$MSE\left(\mathrm{t}_{5(2)}\right) \approx \left[\theta_2\left(\mathrm{S_y} - \mathrm{S_{yx} S_x^{-1} S_{xy}}\right) + \theta_1\left(\mathrm{S_{yx} S_x^{-1} S_{xy}} - \mathrm{S_{yx_{(2)}} S_{22}^{-1} S_{x_{(2)}y}}\right)\right]. \qquad (7.4.9)$$

Comparing (7.4.9) with (7.3.16) we may readily conclude that the estimators $\mathrm{t}_{3(2)}$ and $\mathrm{t}_{5(2)}$ are in the same equivalence class, as both have the same expression for the mean square error matrix.

7.4.3 Multivariate Ratio Estimator for Full Information Case

We have seen earlier that when population information of auxiliary variables is available, then the ratio estimator in two-phase sampling may be constructed by using population information of these auxiliary variables. Hanif et al. (2009) have proposed a multivariate version of the ratio estimator by using population information of auxiliary variables. The proposed estimator is:

$$\mathrm{t}_{6(2)} = \left[\bar{y}_{(1)2}\prod_{j=1}^{q}\left(\frac{\bar{X}_j}{\bar{x}_{(j)2}}\right)^{\alpha_{1j}} \quad \bar{y}_{(2)2}\prod_{j=1}^{q}\left(\frac{\bar{X}_j}{\bar{x}_{(j)2}}\right)^{\alpha_{2j}} \quad \cdots \quad \bar{y}_{(p)2}\prod_{j=1}^{q}\left(\frac{\bar{X}_j}{\bar{x}_{(j)2}}\right)^{\alpha_{pj}}\right]^{/}. \qquad (7.4.10)$$

Hanif et al. (2009) have derived the mean square error of (7.4.10) as below. Using $\bar{y}_{(j)2} = \bar{Y}_j + \bar{e}_{y_{(j)2}}$ etc. in $t_{6(2)}$ and expanding with linear terms only, we may write (7.4.10) as:

$$t_{6(2)} = \begin{bmatrix} \left(\bar{Y}_1 + \bar{e}_{y_{(2)2}}\right)\left(1 - \sum_{j=1}^{q} \frac{\alpha_{1j}}{\bar{X}_j} \bar{e}_{x_{(j)2}}\right) \\ \left(\bar{Y}_2 + \bar{e}_{y_{(2)2}}\right)\left(1 - \sum_{j=1}^{q} \frac{\alpha_{2j}}{\bar{X}_j} \bar{e}_{x_{(j)2}}\right) \\ \vdots \\ \vdots \\ \left(\bar{Y}_1 + \bar{e}_{y_{(p)2}}\right)\left(1 - \sum_{j=1}^{q} \frac{\alpha_{pj}}{\bar{X}_j} \bar{e}_{x_{(j)2}}\right) \end{bmatrix}.$$

The above expression may also be written as:

$$t_{6(2)} = \begin{bmatrix} \left(\bar{Y}_1 + \bar{e}_{y_{(1)2}}\right) \\ \left(\bar{Y}_2 + \bar{e}_{y_{(2)2}}\right) \\ \vdots \\ \vdots \\ \left(\bar{Y}_1 + \bar{e}_{y_{(p)2}}\right) \end{bmatrix} - \begin{bmatrix} \bar{Y}_1 \sum_{j=1}^{q} \frac{\alpha_{1j}}{\bar{X}_j} \bar{e}_{x_{(j)2}} \\ \bar{Y}_2 \sum_{j=1}^{q} \frac{\alpha_{2j}}{\bar{X}_j} \bar{e}_{x_{(j)2}} \\ \vdots \\ \vdots \\ \bar{Y}_p \sum_{j=1}^{q} \frac{\alpha_{pj}}{\bar{X}_j} \bar{e}_{x_{(j)2}} \end{bmatrix}.$$

Using matrix representation, the above expression may be written as:

$$t_{6(2)} - \bar{Y} = \bar{e}_y - A\bar{e}_{x_2}, \qquad (7.4.11)$$

where $A_{(p \times q)} = \left[\left(\bar{Y}_i / \bar{X}_j\right)\alpha_{ij}\right]$ is a matrix containing coefficients. Squaring and applying expectation, the mean square error matrix of $t_{6(2)}$ is:

$$MSE\left(t_{6(2)}\right) \approx \theta_2 S_y - \theta_2 S_{yx} A' - \theta_2 A S_{xy} + \theta_2 A S_x A'. \qquad (7.4.12)$$

Hanif et al. (2009) have shown that the optimum value of **A** which minimizes (7.4.12) is $A = S_{yx}S_x^{-1}$. Substituting the optimum value of **A** in (7.4.12) and simplifying, the mean square matrix of $t_{6(2)}$ is:

$$MSE\left(t_{6(2)}\right) \approx \theta_2\left(S_y - S_{yx}S_x^{-1}S_{xy}\right). \tag{7.4.13}$$

Comparing (7.4.13) with (7.3.7), we may conclude that the estimators $t_{2(2)}$ and $t_{6(2)}$ are in the same equivalence class, as both have the same mean square error matrix.

SECTION-IV:

SINGLE AND TWO PHASE SAMPLING WITH AUXILIARY ATTRIBUTES

CHAPTER EIGHT

USE OF AUXILIARY ATTRIBUTES IN SINGLE PHASE AND TWO-PHASE SAMPLING

8.1 Introduction

Survey statisticians have been using auxiliary information to increase the precision of estimates. Quantitative auxiliary variables had played a very significant role in the development of estimators from time to time in survey sampling. We have elaborated on the role of quantitative auxiliary variables in the development of estimators in previous chapters. The availability of quantitative auxiliary variables is sometimes very difficult and hence the classical ratio and regression estimators are not applicable to estimate population characteristics. As an example, suppose that a survey is initiated to estimate the average weight of inhabitants by using the information about use of a food supplement. The auxiliary information in this example cannot be numerically measured and so the conventional ratio and regression estimators cannot be used for estimation purpose. Fortunately, survey statisticians have developed methods of estimation which incorporate qualitative auxiliary variables. The first known work where qualitative auxiliary variables have been used for estimation is of Naik and Gupta (1996). The authors proposed ratio, product and regression estimators by using information on qualitative auxiliary attributes. Naik and Gupta (1996) observed that their proposed estimators are in line with the conventional ratio, product and regression estimators but with the exception that conventional correlations are replaced with the bi–serial correlations. The work of Naik and Gupta (1996) were extended by Jhajj et al. (2006) when they proposed a family of estimators for single and two-phase sampling by using information of a single auxiliary attribute. Jhajj et al. (2006) showed that a number of estimators emerge from their proposed family as a special case. Shabbir and Gupta (2007) used information on two auxiliary variables and proposed an estimator by using the method of Roy (2003).

The development of estimators by using auxiliary attributes continued and many new estimators have been proposed by a number of survey statisticians. In this chapter we will reproduce the existing estimators for single and two-phase sampling which uses information on one and two qualitative auxiliary variables. We will first discuss the estimators for single phase sampling and then for two-phase sampling.

8.2 Estimators for Single Phase Sampling

Suppose the information on response variable *Y* and a single auxiliary attribute is obtained from a sample of size *n*. The available information can be utilizes to construct several estimators for estimation of population mean or total of *Y*. We know that mean (or total) of the quantitative auxiliary variable plays a key role in building up the estimators for estimation of population characteristics about *Y*. When information about the qualitative auxiliary variable is available, then the mean (or total) of the auxiliary variable cannot be computed, but fortunately the sample and population proportion of auxiliary variable can provide a basis for construction of estimators for estimation of population parameters. The authors who proposed estimators for population characteristics use sample and population proportion of auxiliary attributes to construct the estimators. For notational purpose we will use p to indicate sample proportion and P to indicate population proportion of auxiliary attributes. We will also add the subscripts to p and P when we discus two phase sampling and/or multiple auxiliary attributes case. Specifically the notation $p_{j(h)}$ will be used to indicate sample proportion of *j*-th auxiliary variable at *h*-th phase. For a single auxiliary attribute the subscript outside the bracket will be eliminated and for single phase sampling we will eliminate the subscript within the bracket. Similar notation will be used to indicate the population value of an auxiliary attribute. In the following section we will discuss some of the popular estimators for single phase sampling, while estimators of two-phase sampling will be described in the next section.

8.2.1 Estimators by Naik and Gupta (1996)

Suppose that information of a single auxiliary attribute is available for a sample of size n drawn from a population of size N. Suppose further that the sample proportion of auxiliary attribute is p and corresponding population proportion is P. Using these notations Naik and Gupta (1996) proposed ratio, product and regression estimators using single auxiliary attribute for single phase sampling.

The ratio estimator proposed by Naik and Gupta (1996) was parallel to the conventional ratio estimator but with the exception that sample and population mean of the auxiliary variable were replaced by sample and population proportion of the auxiliary attribute.

8.2.1.1 Ratio Estimator by Naik and Gupta (1996)

The ratio estimator proposed by Naik and Gupta (1996) by using a single auxiliary attribute is:

$$t_{a1(1)} = \bar{y}\frac{P}{p}. \tag{8.2.1}$$

Using the notations from chapter 1, Naik and Gupta (1996) showed that the mean square error of (8.2.1) can be put in the form:

$$MSE\left(t_{a1(1)}\right) \approx E\left(\bar{e}_y - \frac{\bar{Y}}{P}\bar{e}_\tau\right)^2. \tag{8.2.2}$$

Expanding (8.2.2) and applying expectation, Naik and Gupta (1996) have shown that the mean square error of (8.2.1) is:

$$MSE\left(t_{a1(1)}\right) = \theta\bar{Y}^2\left[C_y^2 + C_\tau^2 - 2\rho_{Pb}C_yC_\tau\right]. \tag{8.2.3}$$

In (8.2.3), C_τ^2 is the squared coefficient of variation for the auxiliary variable and ρ_{Pb} is the point bi–serial correlation coefficient between Y and the auxiliary attribute. The mean square error given in (8.2.3) is the same as for the conventional ratio estimator. The only difference is that the mean square error of the conventional ratio estimator is based upon the simple correlation coefficient.

8.2.1.2 Product Estimator by Naik and Gupta (1996)

The product estimator is similar to the ratio estimator but is used when the relationship between the response variable and auxiliary variable is negative. Naik and Gupta (1996) proposed the following product estimator when information of a single auxiliary attribute was available for single phase sampling:

$$t_{a2(1)} = \bar{y}\frac{p}{P}. \tag{8.2.4}$$

The mean square error of $t_{a2(1)}$ can be written as:

$$MSE\left(t_{a2(1)}\right) \approx E\left(\bar{e}_y + \frac{\bar{Y}}{P}\bar{e}_\tau\right)^2. \tag{8.2.5}$$

Expanding (8.2.5) and applying expectation, the mean square error of (8.2.4) is:

$$MSE\left(t_{a2(1)}\right) \approx \theta \bar{Y}^2\left[C_y^2 + C_\tau^2 + 2\rho_{Pb}C_yC_\tau\right]. \quad (8.2.6)$$

The mean square error (8.2.6) is again parallel to the mean square error of the conventional product estimator.

8.2.1.3 Regression Estimator by Naik and Gupta (1996)

Naik and Gupta (1996) also proposed the following regression estimator by using information of a single auxiliary attribute for single phase sampling:

$$t_{a3(1)} = \bar{y} + b\left(p - P\right). \quad (8.2.7)$$

The mean square error of $t_{a3(1)}$ is:

$$MSE\left(t_{a3(1)}\right) \approx \theta\left[\bar{Y}^2C_y^2 + b^2P^2C_\tau^2 + 2bP\bar{Y}C_yC_\tau\rho_{Pb}\right]. \quad (8.2.8)$$

It can be shown that the optimum value of b which minimizes (8.2.8) is:

$$b = \frac{-\bar{Y}C_y}{PC_\tau}\rho_{pb}$$

Using the above value in (8.2.8) and simplifying, we get:

$$MSE\left(t_{a3(1)}\right) = \theta\left(1 - \rho_{Pb}^2\right)\bar{Y}^2C_y^2. \quad (8.2.9)$$

The mean square error (8.2.9) is the same as the mean square error of the conventional regression estimator. The only difference is that the conventional correlation coefficient is replaced by the point bi–serial correlation coefficient.

8.2.2 Regression-Cum-Ratio Estimator by Shabbir and Gupta (2007)

Shabbir and Gupta (2007) proposed the following regression–in–ratio estimator for single phase sampling when information of a single auxiliary attribute is available.

$$t_{a4(1)} = \bar{y}\left[d_1 + d_2\left(P - p\right)\right]\frac{P}{p}, \quad for \quad p > 0 \quad (8.2.10)$$

where d_1 and d_2 are unknown parameters to be obtained so that the mean square error of (8.2.10) is minimum.

Using the notations from chapter 1, the estimator $t_{a4(1)}$ may be written as:

$$t_{a4(1)} = \left(\bar{Y} + \bar{e}_y\right)\left[d_1 + d_2\left(P - P - \bar{e}_\tau\right)\right]\frac{P}{\left(P + \bar{e}_\tau\right)} \tag{8.2.11}$$

Expanding (8.2.11), squaring and applying expectation, the mean square error of $t_{a4(1)}$ is

$$MSE\left(t_{a4(1)}\right) \approx \bar{Y}^2\left[1 + d_1^2 A^* + d_2^2 B^* + 2d_1 d_2 C^* - 2d_1 D^* - 2d_2 E^*\right], \tag{8.2.12}$$

where $A^* = \left[1 + \theta\left\{C_y^2 + C_\tau^2\left(3 - 4k_p\right)\right\}\right]$, $B^* = \theta P^2 C_\tau^2$, $C^* = 2\theta P C_\tau^2\left(1 - k_p\right)$, $D^* = \left[1 + \theta C_\tau^2\left(1 - k_p\right)\right]$, $E^* = \theta P C_\tau^2\left(1 - k_p\right)$ and $k_p = \rho_{pb} C_y / C_\tau$.

The optimum values of d_1 and d_2 which minimize (8.2.12) are:

$$d_1 = \left(B^* D^* - C^* E^*\right)\left(A^* B^* - C^{*2}\right)^{-1}, \tag{8.2.13}$$

$$d_2 = \left(A^* E^* - C^* D^*\right)\left(A^* B^* - C^{*2}\right)^{-1}. \tag{8.2.14}$$

Using (8.2.13) and (8.2.14) in (8.2.12) and on simplification, we get:

$$MSE\left(t_{a4(1)}\right) \approx \bar{Y}^2\left[1 - \frac{\left(B^* D^{*2} - 2C^* D^* E^* + A^* E^{*2}\right)}{\left(A^* B^* - C^{*2}\right)}\right]. \tag{8.2.15}$$

The expression of mean square error given in (8.2.15) has been obtained by Singh and Solanki (2010). Singh and Solanki (2010) have also argued that the mean square expression of (8.2.10) obtained by Shabbir and Gupta (2007) is not the correct one.

8.2.3 Correction in Estimator by Shabbir and Gupta (2007)

$$t_{a5(1)} = \bar{y}\left[d_1 + d_2\left(P_\varphi - p_\varphi\right)\right]\frac{P_\varphi}{p_\varphi} \tag{8.2.16}$$

Using notations from chapter 1, the estimator $t_{a5(1)}$ may be written as:

$$t_{a5(1)} = \left(\bar{Y} + \bar{e}_y\right)\left[d_1 + d_2\left(-\bar{e}_{\tau_\varphi}\right)\right]\frac{P_\varphi}{P_\varphi + \bar{e}_{\tau_\varphi}}$$

$$= \left(\bar{Y} + \bar{e}_y\right)\left[d_1 - d_2\bar{e}_\tau\right]\left(1 + \frac{\bar{e}_{\tau_\varphi}}{P}\right)^{-1}$$

$$t_{a5(1)} \approx \left[d_1\bar{Y} - d_2\bar{Y}\frac{\bar{e}_\tau}{P} - d_2\bar{Y}\,\bar{e}_\tau + d_1\bar{e}_y\right]$$

$$t_{a5(1)} - \bar{Y} \approx \left[\left(d_1 - 1\right)\bar{Y} + d_1\bar{e}_y - d_1\bar{Y}\frac{\bar{e}_\tau}{P} - d_2\bar{Y}\,\bar{e}_\tau\right] \quad (8.2.17)$$

Squaring and taking expectations of (8.2.17)

$$MSE\left(t_{a5(1)}\right) = E\Bigg[\left(d_1 - 1\right)^2\bar{Y}^2 + d_1^2\bar{e}_y^2 + d_1^2\bar{Y}^2\frac{\bar{e}_\tau^2}{P^2} + d_2^2\bar{Y}^2\bar{e}_\tau^2$$
$$+2d_1\left(d_1 - 1\right)\bar{Y}\,\bar{e}_y - 2d_1\left(d_1 - 1\right)\frac{\bar{Y}^2}{P^2}\bar{e}_\tau$$
$$-2d_1\left(d_1 - 1\right)\bar{Y}^2\bar{e}_\tau - 2d_1^2\bar{Y}\,\bar{e}_y\frac{\bar{e}_\tau}{P}$$
$$-2d_1d_2\bar{Y}\,\bar{e}_y\bar{e}_\tau + 2d_1d_2\bar{Y}^2\frac{\bar{e}_\tau^2}{P}\Bigg]$$

$$= \bar{Y}^2\Big[\left(d_1 - 1\right)^2 + d_1^2\theta C_y^2 + \theta d_1^2C_\tau^2 + d_2^2\theta P^2C_\tau^2$$
$$-2d_1^2\theta C_yC_\tau P_{pb} - 2d_1d_2\theta PC_yC_\tau P_{pb} + 2d_1d_2\theta PC_\tau^2\Big] \quad (8.2.18)$$

Differentiating (8.2.18) w.r.t. d_1 and d_2 and on simplifying:

$$d_1 = \frac{1}{1 + \theta\left(1 - P_{pb}^2\right)C_y^2} \quad (8.2.19)$$

$$d_2 = \frac{\left[C_yP_{pb} - C_\tau\right]}{\left[1 + \theta\left(1 - P_{pb}^2\right)C_y^2\right]S_p} \quad (8.2.20)$$

Using (8.2.19) and (8.2.20) in (8.2.18) and on simplifying we get:

$$MSE\left(t_{a5(1)}\right) = \frac{\theta\left(1 - P_{pb}^2\right)S_y^2}{1 + \theta\left(1 - P_{pb}^2\right)\bar{Y}^{-2}S_y^2}$$

8.2.3 Regression-Cum-Ratio Estimator by Singh and Solanki (2010)

Singh and Solanki (2010) constructed the following class of estimators for population mean $\overline{Y}$ as:

$$t_{a6(1)}=\overline{y}\left[d_1+d_2\left(P-\hat{P}\right)\right]\left(\frac{\psi P+\delta\eta}{\psi\hat{P}+\delta\eta}\right)^{\alpha} \tag{8.2.21}$$

where ψ and η are either real numbers or the function of the unknown parameter of auxiliary attribute ϕ such as C_p, $\beta_2(\phi)\rho_{pb}$ and k_p, α is a scalar which takes values -1 (for product-type estimator) and +1 (for ratio-type estimator), 'δ' is an integer which takes values +1 and -1 for designing the estimators such that $\left(\psi\hat{P}+\delta\eta\right)$ and $\left(\psi P+\delta\eta\right)$ are non-negative, and $\left(d_1, d_2\right)$ are to be determined such that MSE of 't' is minimum.

The estimator 't' reduces to the following set of known estimators:

i) For $\left(d_1, d_2, \alpha\right)=\left(1,0,0\right), t=\overline{y}$ usual unbiased estimator,

ii) For $\left(\psi,\eta,\alpha,\delta\right)=\left(1,0,1,1\right), t=\overline{y}\left[d_1+d_2\left(P-\hat{P}\right)\right]\left(P/\hat{P}\right)$

Shabbir and Gupta (2007) estimator,

iii) For $\left(d_1, d_2,\alpha,\delta\right)=\left(1,0,1,1\right), t=\overline{y}\left\{\left(\psi P+\eta\right)/\left(\psi\hat{P}+\eta\right)\right\}$

Abd-Elfattah et al. (2010) estimator.

Many other estimators can be generated from the proposed class of estimators 't' for different suitable values of constants $\left(d_1, d_2,\psi, \eta,\delta,\alpha\right)$.

$$MSE\left(t_{a6(1)}\right)=\overline{Y}^2\left[1+d_1^2A+d_2^2B+2d_1d_2C-2d_1D-2d_2E\right] \tag{8.2.22}$$

where

$$A=\left[1+\left\{\left(1-f\right)/n\right\}\left\{C_y^2+\alpha\nu C_p^2\left(\left(2\alpha+1\right)\nu-4k_p\right)\right\}\right],$$

$$B=\left\{\left(1-f\right)/n\right\}P^2C_p^2,$$

$$C=\left\{2\left(1-f\right)/n\right\}PC_p^2\left(\alpha\nu-k_p\right),$$

$$D=\left[1+\{(1\text{-}f)/n\}\alpha v C_p^2\{\{(\alpha+1)/2\}v\text{-}k_p\}\right],$$

and

$$E=\{(1\text{-}f)/n\}PC_p^2(\alpha v\text{-}k_p)$$

The $MSE\left(t_{a6(1)}\right)$ at (2.7) is minimized for

$$\left.\begin{aligned} d_1&=\{(BD\text{-}CE)/(AB\text{-}C^2)\}=d_1^*\ (\text{say}) \\ d_2&=\{(AE\text{-}CD)/(AB\text{-}C^2)\}=d_2^*\ (\text{say}) \end{aligned}\right\} \quad (8.2.23)$$

Thus the resulting minimum MSE of '$t_{a6(1)}$' is given by

$$MSE_{\min}\left(t_{a6(1)}\right)=\bar{Y}^2\left[1\text{-}\frac{\left(BD^2\text{-}2CDE+AE^2\right)}{\left(AB\text{-}C^2\right)}\right] \quad (8.2.24)$$

8.2.4 Family of Estimators by Jhajj, et al. (2006)

Jhajj et al. (2006) extended the use of the auxiliary attribute in single phase sampling when they proposed a general family of estimators using information on a single auxiliary attribute. The authors proposed a family of estimators and generated a number of estimators as special cases under certain conditions. Jhajj et al. (2006) proposed the following general function as an estimator of population mean in single phase sampling:

$$t_{a7(1)}=g_{\omega}(\bar{y},v)\ for\ v=\frac{p}{P}>0, \quad (8.2.25)$$

where $g_{\omega}(\bar{y},v)$ is any parametric function of $\bar{y}$ and v such that $g_{\omega}(\bar{Y},1)=\bar{Y}$, for all $\bar{Y}$ and satisfy following conditions:

i) Whatever sample is chosen, the point $(\bar{y},v)$ assumes value in a bounded closed convex subset R_2 of two-dimensional real space containing the point $(\bar{Y},1)$,

ii) The function $g_{\omega}(\bar{y},v)$ is continuous and bounded in R_2,

and

iii) The first and second order partial derivatives of $g_{\omega}(\bar{y},v)$ exist and are continuous as well as bounded in R_2.

Some functions that may be possible members of the family of estimators proposed by Jhajj et al. (2006) are:

i) $g_{\omega}\left(\bar{y},v\right)=\bar{y}\,v^{\alpha}$, (8.2.26)

$g_{\omega}\left(\bar{y},v\right)=\bar{y}\,e^{\alpha(v-1)}$, (8.2.27)

ii) $g_{\omega}\left(\bar{y},v\right)=\bar{y}+\alpha\left(v-1\right)$, (8.2.28)

iii) $g_{\omega}\left(\bar{y},v\right)=\bar{y}\,v^{\alpha/2}$, (8.2.29)

iv) $g_{\omega}\left(\bar{y},v\right)=\bar{y}\,v^{\alpha}e^{\alpha(v-1)}$, (8.2.30)

and

v) $g_{\omega}\left(\bar{y},v\right)=\bar{y}\left[k\,v^{\alpha}+\left(1-k\right)\,e^{\alpha(v-1)}\right]$, (8.2.31)

where $0<k<1$, and α is unknown parameter.

Using section 1.4 and expanding (8.2.27) as a second order Taylor's series around the point $\left(\bar{Y},\,1\right)$ we have:

$$t_{a7(1)}=\bar{Y}+\left(\bar{y}-\bar{Y}\right)g_{1w}+\left(v-1\right)g_{2w}$$
$$+\left\{\left(\bar{y}-\bar{Y}\right)^{2}g_{11w}+\left(v-1\right)^{2}g_{22w}+2\left(\bar{y}-\bar{Y}\right)\left(v-1\right)g_{12w}\right\}\Big/2+\cdots \quad (8.2.32)$$

where g_{1a} and g_{2a} are first order derivatives of (8.2.25) w.r.t. $\bar{y}$ and v. Further g_{ija} denotes the corresponding second order derivative of $g_{\omega}\left(\bar{y},v\right)$ at the point $\left(\bar{Y}^{*},v^{*}\right)$ with $\bar{Y}^{*}=\bar{Y}+\theta\left(\bar{y}-\bar{Y}\right)$ and $v^{*}=1+\theta\left(v-1\right)$. Squaring, applying expectation and differentiating, we can see that the mean square error of (8.2.25) is minimized for $g_{2a(opt)}=S_{YP}^{2}\big/S_{P}^{2}$. The mean square error of (8.2.25) is given as:

$$MSE\left(t_{a7(1)}\right)\approx\theta\bar{Y}^{2}C_{Y}^{2}\left(1-\rho_{Pb}^{2}\right). \quad (8.2.33)$$

Jhajj et al. (2006) have further shown that any member of the family has the same mean square error given in (8.2.33). Jhajj et al. (2006) have also obtained an estimate of (8.2.33), i.e.:

$$M\hat{S}E\left(t_{a7(1)}\right)\approx\theta s_{Y}^{2}\left(1-r_{Pb}^{2}\right). \quad (8.2.34)$$

8.2.5 Family of Estimators by Hanif, Haq and Shahbaz (2009)

The family of estimators proposed by Jhajj et al. (2006) used information on a single attributes. Hanif et al. (2009) extended the family of estimators proposed by Jhajj et al. (2006) by using information on two auxiliary attributes. The expression for the family proposed by Hanif et al. (2009) is:

$$t_{a8(1)} = g_w\left(\bar{y}, v_1, v_2\right), \tag{8.2.35}$$

where $v_1 = \frac{p_1}{P_1}, v_2 = \frac{p_2}{P_2}, v_1 > 0, v_2 > 0$, p_1 and p_2 are sample proportions possessing attributes τ_1 and τ_2 respectively, $g_w\left(\bar{y}, v_1, v_2\right)$ is the parametric function such that $g_\omega\left(\bar{Y},1,1\right) = \bar{Y}$, and satisfying the point $\left(\bar{y}, v_1, v_2\right)$ to be in a bounded set in R_3 containing a point $\left(\bar{Y},1,1\right)$. Any function that satisfies $g_w\left(\bar{Y},1,1\right) = \bar{Y}$ can be used as a member of the family (8.2.35). Some possible members of the family of estimators proposed by Hanif et al. (2009) are:

i) $$g_w\left(\bar{y}, v_1, v_2\right) = \bar{y}\, v_1^{\alpha_1} v_2^{\alpha_2}, \tag{8.2.36}$$

ii) $$g_w\left(\bar{y}, v_1, v_2\right) = \bar{y}\, e^{\alpha_1(v_1-1)+\alpha_2(v_2-1)}, \tag{8.2.37}$$

iii) $$g_w\left(\bar{y}, v_1, v_2\right) = \bar{y}\left(v_1 e^{(v_1-1)}\right)^{\alpha_1}\left(v_2 e^{(v_2-1)}\right)^{\alpha_2}, \tag{8.2.38}$$

iv) $$g_w\left(\bar{y}, v_1, v_2\right) = \bar{y} v_1^{\alpha_1} e^{\alpha_2(v_2-1)}, \tag{8.2.39}$$

v) $$g_w\left(\bar{y}, v_1, v_2\right) = \frac{\bar{y}}{2}\left[v_1^{\alpha_1} v_2^{\alpha_2} + e^{\alpha_1(v_1-1)+\alpha_2(v_2-1)}\right], \tag{8.2.40}$$

vi) $$g_w\left(\bar{y}, v_1, v_2\right) = \bar{y} + \alpha_1\left(v_1^{\alpha_3} - 1\right) + \alpha_2\left(v_2^{\alpha_4} - 1\right), \tag{8.2.41}$$

vii) $$g_w\left(\bar{y}, v_1, v_2\right) = \bar{y} + \alpha_1\left(v_1^{\alpha_3} - 1\right) + \alpha_2\left(v_2 - 1\right), \tag{8.2.42}$$

viii) $$g_w\left(\bar{y}, v_1, v_2\right) = \frac{\bar{y}}{k_1 + k_2}\left[k_1 v_1^{\alpha_1/2} + k_2\, e^{\alpha_2(v_2-1)}\right], \tag{8.2.43}$$

ix) $$g_w\left(\bar{y}, v_1, v_2\right) = \bar{y}\left[k\, e^{\alpha_1(v_1-1)} + (1-k)e^{\alpha_2(v_2-1)}\right], \tag{8.2.44}$$

and

x) $$g_w\left(\bar{y}, v_1, v_2\right) = \bar{y} + \alpha_1\left(v_1 - 1\right) + \alpha_2\left(v_2 - 1\right), \tag{8.2.45}$$

where k_1 and k_2 are positive numbers and $0 < k < 1$.

Using section 1.4 in (8.2.35), expanding using second order Taylor's expansion we have:

$$\begin{aligned}t_{a8(1)} &= \bar{Y} + (\bar{y} - \bar{Y}) g_{1w} + (v_1 - 1) g_{2w} + (v_2 - 1) g_{3w} \\ &+ \frac{1}{2}\Big\{(\bar{y} - \bar{Y})^2 g_{11w} + (v_1 - 1)^2 g_{22w} + (v_2 - 1)^2 g_{33w} \\ &+ 2(\bar{y} - \bar{Y})(v_1 - 1) g_{12w} + 2(\bar{y} - \bar{Y})(v_2 - 1) g_{13w} \\ &+ 2(v_1 - 1)(v_2 - 1) g_{23w}\Big\} + \cdots\end{aligned} \quad (8.2.46)$$

where $g_{iw}; i = 1,2,3;$ are first order derivatives of $t_{a8(1)}$ w.r.t. $(\bar{y}, v_1, v_2)$ and g_{ijw}; $(i, j) = 1,2,3; i \neq j;$ are second order derivatives at the points $(\bar{Y}^*, v_1^*, v_2^*)$ with $\bar{Y}^* = \bar{Y} + \theta(\bar{y} - \bar{Y}), v_1^* = 1 + \theta(v_1 - 1)$ and $v_2^* = 1 + \theta(v_2 - 1)$. Applying the expectation we can readily see that $E(t_{a8(1)}) = \bar{Y} + O(n^{-1})$. Squaring and applying expectation we have:

$$MSE(t_{a8(1)}) \approx \theta \bar{Y}^2 C_y^2 (1 - \rho_{y.\tau_1\tau_2}^2). \quad (8.2.47)$$

Hanif et al. (2009) have shown that any member of the family (8.2.35) will have the mean square error given in (8.2.47). For the purpose of illustration, consider the estimator (8.2.36) as:

$$g_w(\bar{y}, v_1, v_2) = \bar{y}\, v_1^{\alpha_1} v_2^{\alpha_2}. \quad (8.2.48)$$

Using the values of v_1 and v_2 in (8.2.36) we have:

$$g_\omega(\bar{y}, v_1, v_2) = (\bar{Y} + \bar{e}_y)(1 + \bar{e}_{\tau_1}/\mathrm{P}_1)^{\alpha_1} (1 + \bar{e}_{\tau_2}/\mathrm{P}_2)^{\alpha_2} \quad (8.2.49)$$

Expanding (8.2.48), squaring and applying expectation we have:

$$MSE(g_\omega(\bar{y}, v_1, v_2)) \approx E\left[\bar{e}_y + \frac{\alpha_1 \bar{Y}}{\mathrm{P}_1}\bar{e}_{\tau_1} + \frac{\alpha_2 \bar{Y}}{\mathrm{P}_2}\bar{e}_{\tau_2}\right]$$

On simplification, we can see that the mean square error of (8.2.36) can be written as:

$$\begin{aligned}MSE(g_w(\bar{y}, v_1, v_2)) \approx \theta\Big[&\bar{Y}^2 C_y^2 + \alpha_1^2 \bar{Y}^2 C_{\tau_1}^2 + \alpha_2^2 \bar{Y}^2 C_{\tau_2}^2 \\ &+ 2\alpha_1 \bar{Y}^2 C_y C_{\tau_1} \rho_{Pb_1} + 2\alpha_2 \bar{Y}^2 C_y C_{\tau_2} \rho_{Pb_2} \\ &+ 2\alpha_1\alpha_2 \bar{Y}^2 C_{\tau_1} C_{\tau_2} Q_{12}\Big].\end{aligned} \quad (8.2.50)$$

The optimum values of α_1 and α_2 which minimize (8.2.50) are:

$$\alpha_1 = \frac{-C_y\left(\rho_{Pb_1} - \rho_{Pb_2}Q_{12}\right)}{C_{\tau_1}\left(1-Q_{12}^2\right)}, \qquad (8.2.51)$$

and $$\alpha_2 = \frac{-C_y\left(\rho_{Pb_2} - \rho_{Pb_1}Q_{12}\right)}{C_{\tau_2}\left(1-Q_{12}^2\right)}. \qquad (8.2.52)$$

Using (8.2.51) and (8.2.52) in (8.2.50) and simplifying, the mean square error of (8.2.36) is:

$$MSE\left(g_\omega\left(\bar{y}, v_1, v_2\right)\right) \approx \theta\left(1\text{-}\rho_{y.\tau_1\tau_2}^2\right)\bar{Y}^2 C_y^2, \qquad (8.2.53)$$

which is (8.2.47). The estimate of (8.2.47) is written as:

$$M\hat{S}E\left(t_{a8(1)}\right) \approx \theta\bar{y}^2 c_Y^2\left(1-r_{y.\tau_1\tau_2}^2\right) \qquad (8.2.54)$$

Any member of the family and (8.2.54) provide the basis for constructing the confidence interval for population mean.

8.3 Estimators for Two-Phase Sampling

The use of an auxiliary attribute in single phase sampling has been explored in section 8.2. We have reported some popular estimators which utilize information of single or two auxiliary attributes in single phase sampling. Many survey statisticians have used the auxiliary attributes to propose the estimators for two-phase sampling. The idea used in developing these estimators goes parallel to conventional ideas of the use of an auxiliary variable in two phase sampling. Various authors have proposed various estimators using information on single and two auxiliary attributes in two phase sampling from time to time. We present some of the well-known estimators for two phase sampling which use information on single and two auxiliary attributes. These estimators are discussed below.

8.3.1 Regression-Cum-Ratio Estimator by Shabbir and Gupta (2007)

Shabbir and Gupta (2007) proposed the following regression–cum–ratio type estimator using information of a **single auxiliary** attribute.

$$t_{a1(2)} = \bar{y}_2\left[w_1 + w_2\left(p_{(1)} - p_{(2)}\right)\right]\frac{p_{(1)}}{p_{(2)}}, \quad for \;\; p_{(2)} > 0 \qquad (8.3.1)$$

where $p_{(i)}$ is sample proportion of auxiliary attribute at ith phase. Using section 1.4 and expanding (8.3.1) we have:

$$t_{a1(2)} = \left(\bar{Y} + \bar{e}_{y_2}\right)\left[w_1 + w_2\left(\bar{e}_{\tau_{(2)}} - \bar{e}_{\tau_{(1)}}\right)\right]\frac{\left(1+\bar{e}_{\tau_{(1)}}/P\right)}{\left(1+\bar{e}_{\tau_{(2)}}/P\right)}$$

Squaring and applying expectation, the mean square error of (8.3.1) may be written as:

$$MSE\left(t_{a1(2)}\right) = \bar{Y}^2\left[1 + w_1^2 A_1^* + w_2^2 B_1^* + 2w_1 w_2 C_1^* - 2w_1 D_1^* - 2w_2 E_1^*\right],$$

where $A_1^* = \left[1+\theta C_y^2 + \left(\theta_2 - \theta_1\right)C_\tau^2\left(3-4k_p\right)\right]$, $B_1^* = \left(\theta_2-\theta_1\right)P^2C_\tau^2$

$$C_1^* = 2P\left(\theta_2-\theta_1\right)C_\tau^2\left(1-k_p\right),\ D_1^* = \left[1+\left(\theta_2-\theta_1\right)C_\tau^2\left(1-k_p\right)\right]$$

$$E_1^* = \left(\theta_2-\theta_1\right)PC_\tau^2\left(1-k_p\right) \text{ and } k_p = \rho_{pb}\left(C_y/C_\tau\right) \tag{8.3.2}$$

The optimum value of w_1 and w_2 which minimizes (8.3.2) are:

$$w_1 = \left(B_1^* D_1^* - C_1^* E_1^*\right)\left(A_1^* B_1^* - C_1^{*2}\right)^{-1}$$

$$w_2 = \left(A_1^* E_1^* - C_1^* D_1^*\right)\left(A_1^* B_1^* - C_1^{*2}\right)^{-1}$$

Using W_1 and W_2 in (8.3.2) and simplifying, the mean square error of (8.3.1) is:

$$MSE\left(t_{a1(2)}\right) = \bar{Y}^2\left[1 - \frac{B_1^* D_1^{*2} - C_1^* D_1^* E_1^* + A_1^* E_1^{*2}}{\left(A_1^* B_1^* - C_1^{*2}\right)^{-1}}\right]. \tag{8.3.3}$$

The mean square error given in (8.3.3) is the mean square error of the shrinkage version of any member of the family of estimators proposed by Jhajj et al. (2006) for two-phase sampling, which we present in the following section.

8.3.2 Correction in Estimator by Shabbir and Gupta (2007)

Shabbir and Gupta (2007) proposed the following regression–cum–ratio type estimator using information of a **single auxiliary** attribute.

$$t_{a2(2)} = \bar{y}_2\left[w_1 + w_2\left(p_{(1)} - p_{(2)}\right)\right]\frac{p_{(1)}}{p_{(2)}}, \qquad p_{(2)} > 0$$

(8.3.4)

where $p_{(i)}$ is sample proportion of the auxiliary attribute at ith phase. Using section 1.4 and expanding (8.3.4) we have:

$$t_{a2(2)} = \left(\bar{y} + \bar{e}_{y_2}\right)\left[w_1 + w_2\left(P + \bar{e}_{\tau_{(1)}} - P - \bar{e}_{\tau(2)}\right)\right]\frac{\left(P + \bar{e}_{\tau_{(1)}}\right)}{\left(P + \bar{e}_{\tau_{(2)}}\right)}$$

$$= \left(\bar{y} + \bar{e}_{y_2}\right)\left[w_1 - w_2\left(\bar{e}_{\tau_{(2)}} - \bar{e}_{\tau_{(1)}}\right)\right]\left[1 + \frac{\bar{e}_{\tau_{(1)}}}{P}\right]\left[1 + \frac{\bar{e}_{\tau_{(2)}}}{P}\right]^{-1}$$

On simplification we get:

$$\left(t_{a2(2)} - \bar{Y}\right) \approx \left[(w_1 - 1)\bar{Y} - w_2\left(\bar{e}_{\tau_{(2)}} - \bar{e}_{\tau_{(1)}}\right) - \frac{w_1\bar{Y}}{P}\left(\bar{e}_{\tau_{(2)}} - \bar{e}_{\tau_{(1)}}\right) + w_1\bar{e}_{y_2}\right]$$

Squaring and Taking Expectation:

$$MSE\left(t_{a2(2)}\right) \approx E\begin{bmatrix} (w_1 - 1)^2\bar{Y}^2 + w_2^2\bar{Y}^2\left(\bar{e}_{\tau_{(2)}} - \bar{e}_{\tau_{(1)}}\right)^2 \\ +\frac{w_1^2\bar{Y}^2}{P^2}\left(\bar{e}_{\tau_{(2)}} - \bar{e}_{\tau_{(1)}}\right)^2 + w_1^2\bar{e}_y^2 + 2w_1w_2\frac{\bar{Y}^2}{P}\left(\bar{e}_{\tau_{(2)}} - \bar{e}_{\tau_{(1)}}\right)^2 \\ -2\frac{w_1^2\bar{Y}}{P}\left(\bar{e}_{\tau_{(2)}} - \bar{e}_{\tau_{(1)}}\right)\bar{e}_{y_2} - 2w_1w_2\bar{Y}\left(\bar{e}_{\tau_{(2)}} - \bar{e}_{\tau_{(1)}}\right)\bar{e}_{y_2} \end{bmatrix}$$

or

$$MSE\left(t_{a2(2)}\right) = \bar{Y}^2\Big[(w_1 - 1)^2 + w_1^2\theta_2C_y^2 + (\theta_2 - \theta_1)\big\{w_2^2P^2C_\tau^2 + w_1^2C_\tau^2$$
$$+2w_1w_2PC_\tau^2 - 2w_1^2C_yC_\tau P_{pb} - 2w_1w_2PC_yC_\tau P_{pb}\big\}\Big] \quad (8.3.5)$$

Differentiating (8.3.5) w.r.t w_1 and w_2 we get optimum value of w_1 & w_2 as:

$$w_1 = \frac{1}{1 + \left[\theta_2 - (\theta_2 - \theta_1)P_{pb}^2\right]C_y^2}$$

$$w_2 = \frac{\left[P_{pb}C_y - C_\tau\right]}{\left[1 + \left\{\theta_2 - (\theta_2 - \theta_1)P_{pb}^2\right\}C_y^2\right]S_p}$$

Using optimum value of w_1 and w_2 in (8.3.5) and on simplification we get:

$$MSE\left(t_{a2(2)}\right) = \frac{\left[\theta_2 - (\theta_2 - \theta_1)P_{pb}^2\right]S_y^2}{1 + \left[\theta_2 - (\theta_2 - \theta_1)P_{pb}^2\right]\bar{Y}^{-2}S_y^2} \tag{8.3.6}$$

The mean square error given in (8.3.6) is the mean square error of the shrinkage version of any member of the family of estimators proposed by Jhajj et al. (2006) for two-phase sampling, which we present in the following section.

8.3.3 Family of Estimators by Jhajj, Sharma and Grover (2006) for Single Auxiliary Attribute

Jhajj et al. (2006) proposed a general family of estimators for use in two-phase sampling using information on a **single** auxiliary attribute. The family produces several estimators as special cases.

The general expression for the proposed family of estimators is:

$$t_{a3(2)} = g_{\omega}\left(\bar{y}_2, v_d\right) \tag{8.3.7}$$

where $v_d = p_{(2)}/p_{(1)}$, and $g_{\omega}\left(\bar{Y},1\right) = \bar{Y}$ as before.

Jhajj et al. (2006) suggested that any function which satisfies $g_{\omega}\left(\bar{Y},1\right) = \bar{Y}$ may be used as a member of the family (8.3.7). Some of the possible members of the family (8.3.7) are:

i) $g_{\omega}\left(\bar{y}_2, v_d\right) = \bar{y}_2 v_d^{\alpha}$ (8.3.8)

ii) $g_{\omega}\left(\bar{y}_2, v_d\right) = \bar{y}_2 + \alpha\left(v_d - 1\right)$,

(8.3.9)

iii) $g_{\omega}\left(\bar{y}_2, v_d\right) = \bar{y}_2 + e^{\alpha(v_d - 1)}$, (8.3.10)

and

iv) $g_{\omega}\left(\bar{y}_2, v_d\right) = \bar{y}_2 + e^{\alpha(v_d - 1)} v_d^{\alpha}$, (8.3.11)

where α is an unknown parameter.

Jhajj et al. (2006) have shown that the mean square error of (8.3.7) to $O(1/n)$ is:

$$MSE\left(t_{a3(2)}\right) \approx \left(\theta_2 - \theta_3\rho_{Pb}^2\right)\bar{Y}^2 C_y^2, \tag{8.3.12}$$

The mean square error given in (8.3.12) is the mean square error of the family of estimators given in (8.3.7) and hence any member of the family will have the same mean square error as given in (8.3.12).We illustrate this idea by using the following member of the family:

$$g_\omega\left(\bar{y}_2, v_d\right) = \bar{y}_2\, v_d^\alpha, \qquad (8.3.13)$$

or $g_\omega\left(\bar{y}_2, v_d\right) \approx \left(\bar{Y} + \bar{e}_{y_2}\right)\left\{1 + P^{-1}\left(\bar{e}_{\tau_{1(2)}} - \bar{e}_{\tau_{1(1)}}\right)\right\}^\alpha.$

Expanding, squaring and applying expectation, the mean square error of (8.3.13) is:

$$MSE\left\{g_\omega\left(\bar{y}_2, v_d\right)\right\} \approx E\left[\bar{e}_{y_2}^2 + P^{-2}\alpha^2\bar{Y}^2\left(\bar{e}_{\tau_{(2)}} - \bar{e}_{\tau_{(1)}}\right)^2 + 2\alpha P^{-1}\bar{Y}\bar{e}_{y_2}\left(\bar{e}_{\tau_{(2)}} - \bar{e}_{\tau_{(1)}}\right)\right]$$

Applying expectation we have:

$$MSE\left\{g_\omega\left(\bar{y}_2, v_d\right)\right\} \approx \left[\theta_2\bar{Y}^2C_y^2 + \theta_3\left\{\alpha^2\bar{Y}^2C_\tau^2 + 2\alpha\bar{Y}^2C_yC_\tau\rho_{Pb}\right\}\right]. \qquad (8.3.14)$$

Differentiating (8.3.14) w.r.t. α and equating to zero, we have $\alpha = -C_y\rho_{Pb}/C_\tau$. Using this value of α in (8.3.10) and simplifying, the mean square error of (8.3.13) is:

$$MSE\left\{g_\omega\left(\bar{y}, v\right)\right\} \approx \left(\theta_2 - \theta_3\rho_{Pb}^2\right)\bar{Y}^2C_y^2,$$

which is (8.3.12). We, therefore, have concluded that the mean square error of any member of the family (8.3.4) will be the same as given in (8.3.12). The estimate of mean square error is:

$$\hat{MSE}\left(t_{a3(2)}\right) \approx \left(\theta_2 - \theta_3 r_{Pb}^2\right)s_y^2\,. \qquad (8.3.15)$$

The estimated mean square error given in (8.3.15) along with any member of the family (8.3.7) may be used to construct the confidence interval for population mean.

8.3.4 Families of Estimators by Hanif, et al. (2009) for Two Auxiliary Attributes

Hanif et al. (2009) extended the family of estimators proposed by Jhajj et al. (2006) by using information on **two** auxiliary attributes. Hanif et al.

(2009) have illustrated various possibilities which can arise in two phase sampling when two or more auxiliary attributes are utilized for estimation of population characteristics. Hanif et al. (2009) constructed the estimators using two auxiliary attributes for the following situations:

- Population proportion of none of the auxiliary attributes is known; referred to as the No Information Case (NIC)
- Population proportion of one of the auxiliary attributes is known; referred to as the Partial Information Case (PIC)
- Population proportion of both auxiliary attributes is known; referred to as the Full Information Case (FIC). This eventually leads one toward single phase sampling.

Hanif et al. (2009) proposed a family of estimator for each of the above mentioned situations. We discuss these families of estimators in the following.

8.3.4.1 Family of Estimators for No Information Case

Hanif et al. (2009) have given a further family of estimators for two-phase sampling which has a much wider applicability. This family of estimators has been proposed by assuming that information about auxiliary attributes is available for the second phase sample only. Hanif et al. (2009) have termed this situation as the No Information Case. The family of estimators is:

$$t_{a4(2)} = g_{\omega}\left(\bar{y}_2, v_{1d}, v_{2d}\right), \tag{8.3.16}$$

where $v_{1d} = p_{1(2)}/p_{1(1)}, v_{2d} = p_{2(2)}/p_{2(1)}, v_{1d} > 0, v_{2d} > 0$. Again the same restriction is imposed on $t_{a4(2)} = g_{\omega}\left(\bar{y}_2, v_{1d}, v_{2d}\right)$, that is $g_{\omega}\left(\bar{Y},1,1\right) = \bar{Y}$. Hanif et al. (2009) have suggested some of the possible members of the family (8.3.16) as:

i) $g_{\omega}\left(\bar{y}_2, v_{1d}, v_{2d}\right) = \bar{y}_2\, v_{1d}^{\alpha_1} v_{2d}^{\alpha_2}$,

ii) $g_{\omega}\left(\bar{y}_2, v_{1d}, v_{2d}\right) = \bar{y}_2\, e^{\alpha_1(v_{1d}-1)+\alpha_2(v_{2d}-1)}$,

iii) $g_{\omega}\left(\bar{y}_2, v_{1d}, v_{2d}\right) = \bar{y}_2\left(v_{1d}e^{(v_{1d}-1)}\right)^{\alpha_1}\left(v_{2d}e^{(v_{2d}-1)}\right)^{\alpha_2}$,

iv) $g_{\omega}\left(\bar{y}_2, v_{1d}, v_{2d}\right) = \bar{y}_2 v_{1d}^{\alpha_1}\, e^{\alpha_2(v_{2d}-1)}$,

v) $g_{\omega}\left(\bar{y}_2, v_{1d}, v_{2d}\right)=\frac{\bar{y}_2}{2}\left[v_{1d}^{\alpha_1} v_{2d}^{\alpha_2}+e^{\alpha_1\left(v_{1d}-1\right)+\alpha_2\left(v_{2d}-1\right)}\right]$,

vi) $g_{\omega}\left(\bar{y}_2, v_{1d}, v_{2d}\right)=\bar{y}_2+\alpha_1\left(v_{1d}^{\alpha_3}-1\right)+\alpha_2\left(v_{2d}^{\alpha_4}-1\right)$,

vii) $g_{\omega}\left(\bar{y}_2, v_{1d}, v_{2d}\right)=\bar{y}_2+\alpha_1\left(v_{1d}^{\alpha_3}-1\right)+\alpha_2\left(v_{2d}-1\right)$

viii) $g_{\omega}\left(\bar{y}_2, v_{1d}, v_{2d}\right)=\frac{\bar{y}_2}{k_1+k_2}\left[k_1 v_{1d}^{\alpha_1/2}+k_2\, e^{\alpha_2\left(v_{2d}-1\right)}\right]$,

ix) $g_{\omega}\left(\bar{y}_2, v_{1d}, v_{2d}\right)=\bar{y}_2\left[k\, e^{\alpha_1\left(v_{1d}-1\right)}+(1-k) e^{\alpha_2\left(v_{2d}-1\right)}\right]$,

and

x) $g_{\omega}\left(\bar{y}_2, v_{1d}, v_{2d}\right)=\bar{y}_2+\alpha_1\left(v_{1d}-1\right)+\alpha_2\left(v_{2d}-1\right)$,

where k_1 and k_2 are positive numbers and $0<k<1$.

Following the lines of Hanif et al. (2009), as in section 8.2.4, the mean square error of family (8.3.16) to $O(1/n)$ is:

$$MSE\left(t_{a4(2)}\right) \approx \bar{Y}^2 C_y^2\left[\theta_2\left(1-\rho_{y.\tau_1\tau_2}^2\right)+\theta_1 \rho_{y.\tau_1\tau_2}^2\right], \tag{8.3.17}$$

where $\rho_{y.\tau_1\tau_2}^2$ is the squared multiple point bi–serial correlation coefficient between Y and two auxiliary attributes. Hanif et al. (2009) has further shown that any member of the family (8.3.16) will have the same mean square error as given in (8.3.17). For the purpose of illustration we consider:

$$g_{\omega}\left(\bar{y}_2, v_{1d}, v_{2d}\right)=\bar{y}_2\, v_{1d}^{\alpha_1} v_{2d}^{\alpha_2}.$$

Expanding we have:

$$g_{\omega}\left(\bar{y}_2, v_{1d}, v_{2d}\right)-\bar{Y} \approx\left[\bar{e}_{y_2}+P_1^{-1} \alpha_1 \bar{Y}\left(\bar{e}_{\tau_{1(2)}}-\bar{e}_{\tau_{1(1)}}\right)+P_2^{-1} \alpha_2 \bar{Y}\left(\bar{e}_{\tau_{2(2)}}-\bar{e}_{\tau_{2(1)}}\right)\right].$$

Squaring and applying expectation, the mean square error of (7.3.18) is:

$$MSE\left\{g_{\omega}\left(\bar{y}_2, v_{1d}, v_{2d}\right)\right\} \approx\left[\theta_2 \bar{Y}^2 C_y^2+\theta_3\left\{\begin{array}{l}\alpha_1^2 \bar{Y}^2 C_{\tau_1}^2+\alpha_2^2 \bar{Y}^2 C_{\tau_2}^2+2 \alpha_1 \bar{Y}^2 C_y C_{\tau_1} \rho_{Pb_1} \\ +2 \alpha_2 \bar{Y}^2 C_y C_{\tau_2} \rho_{Pb_2}+2 \alpha_1 \alpha_2 \bar{Y}^2 C_{\tau_1} C_{\tau_2} Q_{12}\end{array}\right\}\right].$$

The optimum value of α_1 and α_2 are the same as given in (8.3.16) and (8.3.17). Using those values in the above equation and simplifying we have:

$$MSE\left\{g_{\omega}\left(\bar{y}_2, v_{1d}, v_{2d}\right)\right\} \approx \left[\theta_2\left(1-\rho^2_{y.\tau_1\tau_2}\right)+\theta_1\rho^2_{y.\tau_1\tau_2}\right]\bar{Y}^2C_y^2,$$

which is (8.3.17).

8.3.4.2 Family of Estimators for Partial Information Case

The family of estimators for the partial information case, proposed by Hanif et al. (2009), is:

$$t_{a5(2)} = g_{\omega}\left(\bar{y}_2, v_1, v_{2d}\right) \tag{8.3.18}$$

where $v_1 = p_{1(2)}/P_1$; $v_{2d} = p_{2(2)}/p_{2(1)}, v_1 > 0, v_{2d} > 0$. Hanif et al. (2009) have argued that any parametric function satisfying $g_{\omega}\left(\bar{Y},1,1\right)=\bar{Y}$ can be used as a member of family (8.3.18). Some possible members of the family (8.3.18) pointed out by Hanif et al. (2009) are:

i) $g_{\omega}\left(\bar{y}_2, v_1, v_{2d}\right) = \bar{y}_2\, v_1^{\lambda_1} v_{2d}^{\lambda_2}$

ii) $g_{\omega}\left(\bar{y}_2, v_1, v_{2d}\right) = \bar{y}_2\, e^{\lambda_1(v_1-1)+\lambda_2(v_{2d}-1)}$

iii) $g_{\omega}\left(\bar{y}_2, v_1, v_{2d}\right) = \bar{y}_2\left(v_1 e^{(v_1-1)}\right)^{\lambda_1}\left(v_{2d}e^{(v_{2d}-1)}\right)^{\lambda_2}$

iv) $g_{\omega}\left(\bar{y}_2, v_1, v_{2d}\right) = \bar{y}_2 v_1^{\lambda_1}\, e^{\lambda_2(v_{2d}-1)}$

v) $g_{\omega}\left(\bar{y}_2, v_1, v_{2d}\right) = \dfrac{\bar{y}_2}{2}\left[v_1^{\lambda_1} v_{2d}^{\lambda_2} + e^{\lambda_1(v_1-1)+\lambda_2(v_{2d}-1)}\right]$

vi) $g_{\omega}\left(\bar{y}_2, v_1, v_{2d}\right) = \bar{y}_2 + \lambda_1\left(v_1^{\alpha_3}-1\right)+\lambda_2\left(v_{2d}^{\alpha_4}-1\right)$

vii) $g_{\omega}\left(\bar{y}_2, v_1, v_{2d}\right) = \bar{y}_2 + \lambda_1\left(v_1^{\alpha_3}-1\right)+\lambda_2\left(v_{2d}-1\right)$

viii) $g_{\omega}\left(\bar{y}_2, v_1, v_{2d}\right) = \dfrac{\bar{y}_2}{k_1+k_2}\left[k_1 v_1^{\frac{\lambda_1}{2}} + k_2\, e^{\lambda_2(v_{2d}-1)}\right]$

ix) $g_{\omega}\left(\bar{y}_2, v_1, v_{2d}\right) = \bar{y}_2\left[k\, e^{\lambda_1(v_1-1)} + (1-k)e^{\lambda_2(v_{2d}-1)}\right]$

and

x) $g_{\omega}\left(\bar{y}_2, v_1, v_{2d}\right) = \bar{y}_2 + \lambda_1\left(v_1-1\right)+\lambda_2\left(v_{2d}-1\right)$

where k_1 and k_2 are positive numbers and $0 < k < 1$.

The mean square error of (8.3.18) can be obtained following the method of Haq and Hanif; some algebra leads us to following mean square error of $t_{a5(2)}$:

$$MSE\left(t_{a5(2)}\right) \approx \theta_2 \bar{Y}^2 C_y^2 \left[1 - \frac{\theta_2 \rho_{Pb_1}^2 + \theta_3 \rho_{Pb_2}^2 - 2\theta_3 Q_{12} \rho_{Pb_1} \rho_{Pb_2}}{\left(\theta_2 - \theta_3 Q_{12}^2\right)}\right], \tag{8.3.19}$$

where Q_{12} is the coefficient of association between two auxiliary variables.

We can show that the mean square error of any member of (8.3.18) is the same and is given in (8.3.19). We justify this with the help of one example below:

Consider a member of (8.3.18) as:

$$g_\omega\left(\bar{y}_2, v_1, v_{2d}\right) = \bar{y}_2 \, v_1^{\lambda_1} v_{2d}^{\lambda_2}.$$

The above estimator can be written as:

$$\left(g_\omega\left(\bar{y}_2, v_1, v_{2d}\right) - \bar{Y}\right) \approx E\left[\bar{e}_{y_2} + P_1^{-1}\lambda_1 \bar{Y} \bar{e}_{\tau_{1(2)}} + P_2^{-1}\lambda_2 \bar{Y}\left(\bar{e}_{\tau_{2(2)}} - \bar{e}_{\tau_{2(1)}}\right)\right].$$

Squaring, applying expectation and simplifying, the mean square error is:

$$MSE\left(g_\omega\left(\bar{y}_2, v_1, v_{2d}\right)\right) \approx \left[\theta_2\left\{\bar{Y}^2 C_y^2 + \lambda_1^2 \bar{Y}^2 C_{\tau_1}^2 + 2\lambda_1 \bar{Y}^2 C_y C_{\tau_1} \rho_{Pb_1}\right\}\right.$$
$$\left. + \theta_3\left\{\lambda_2^2 \bar{Y}^2 C_{\tau_2}^2 + 2\lambda_2 \bar{Y}^2 C_y C_{\tau_2} \rho_{Pb_2} + 2\lambda_1 \lambda_2 \bar{Y}^2 C_{\tau_1} C_{\tau_2} Q_{12}\right\}\right]. \tag{8.3.20}$$

The optimum values of λ_1 and λ_2 that minimize (8.3.20) are:

$$\lambda_1 = \frac{-C_y\left(\theta_2 \rho_{Pb_1} - \rho_{Pb_2} \theta_3 Q_{12}\right)}{C_{\tau_1}\left(\theta_2 - \theta_3 Q_{12}^2\right)}, \tag{8.3.21}$$

and

$$\lambda_2 = \frac{-\theta_2 C_y\left(\rho_{Pb_2} - \rho_{Pb_1} Q_{12}\right)}{C_{\tau_2}\left(\theta_2 - \theta_3 Q_{12}^2\right)}. \tag{8.3.22}$$

Using (8.3.21) and (8.3.22) in (8.3.20) and simplifying, the mean square is:

$$MSE\left(t_{a5(2)}\right)\approx\theta_2\bar{Y}^2C_y^2\left[1-\frac{\theta_2\rho_{Pb_1}^2+\theta_3\rho_{Pb_2}^2-2\theta_3Q_{12}\rho_{Pb_1}\rho_{Pb_2}}{\left(\theta_2-\theta_3Q_{12}^2\right)}\right],$$

which is (8.3.19).

The family of estimators proposed by Hanif et al. (2009) may be slightly modified by defining $v_1=p_{1(1)}/P_1, v_{2d}=p_{2(2)}/p_{2(1)}, v_1>0, v_{2d}>0$. Hanif et al. (2009) have shown that the mean square error of $t_{a5(2)}$ is:

$$MSE\left(t_{a5(2)}\right)\approx\bar{Y}^2C_y^2\left\{\theta_2\left(1-\rho_{pb_2}^2\right)+\theta_3\left(\rho_{pb_2}^2-\rho_{pb_1}^2\right)\right\}. \qquad (8.3.23)$$

Hanif et al. (2009) have further shown that any member of the family (8.2.12) will have the same mean square error as given in (8.3.23).

Shahbaz and Hanif (2009) proposed the following shrinkage version of $t_{a5(2)}$ as estimator:

$$t_{a6(2)}=dt_{a5(2)}=\frac{t_{a5(2)}}{1+\bar{y}^{-2}MSE(t_{a5(2)})} \qquad (8.3.24)$$

Where $t_{a5(2)}$ is any available estimator of parameter $\bar{Y}$ and

$d=\frac{1}{1+\bar{y}^{-2}MSE(t_{a5(2)})}$ is shrinkage constant.

The mean square error of $t_{a6(2)}$ is:

$$MSE(t_{a6(2)})=\frac{MSE(t_{a5(2)})}{1+\bar{y}^{-2}MSE(t_{a5(2)})} \qquad (8.3.25)$$

8.3.5 Regression –Cum-Ratio Estimator by Singh and Solanki (2010)

Singh and Solanki (2010) defined the double sampling version of the proposed class of estimators ' $t_{a7(2)}$ ' for population mean $\bar{Y}$ as:

$$t_{a7(2)}=\bar{y}\left[w_1+w_2\left(\hat{P}'-\hat{P}\right)\right]\left(\frac{\psi\hat{P}'+\delta\eta}{\psi\hat{P}+\delta\eta}\right)^{\alpha} \qquad (8.3.26)$$

where ψ and η are either real numbers or the function of the known parameter of an auxiliary attribute such as C_p, $\beta_2(\phi)\rho_{pb}$ and k_p, α is a scalar which takes values -1 (for product-type estimator) and +1 (for ratio-type estimator), 'δ' is an integer which takes values +1 and -1 for designing the estimators such that $\left(\psi\hat{P}'+\delta\eta\right)$ and $\left(\psi\hat{P}+\delta\eta\right)$ are non-negative, and (w_1, w_2) are constants whose values are to be determined but whose sum is not necessarily equal to one.

To the first degree of the approximation, the bias and MSE of $t_{a7(2)}$ are respectively given by:

$$B(t_{a7(2)})=\bar{Y}\left[(w_1-1)+\left(\frac{1}{n}-\frac{1}{n'}\right)C_p^2\left\{\alpha v\left(\frac{w_1(\alpha+1)v}{2}+w_2P\right)-(w_1\alpha v+w_2P)k_p\right\}\right] \tag{8.3.27}$$

$$\text{MSE}\left(t_{a7(2)}\right)=\bar{Y}^2\left[1+w_1^2A_1+w_2^2B_1+1w_1w_2C_1-2w_1D_1-2w_2E_1\right] \tag{8.3.28}$$

where

$$A_1=\left[1+\frac{(1-f)}{n}C_y^2+\left(\frac{1}{n}-\frac{1}{n'}\right)\alpha v C_p^2\left\{(2\alpha+1)v-4k_p\right\}\right]$$

$$B_1=\left(\frac{1}{n}-\frac{1}{n'}\right)P^2C_p^2$$

$$C_1=2P\left(\frac{1}{n}-\frac{1}{n'}\right)\left(\alpha v-k_p\right)C_p^2$$

$$D_1=\left[1+\left(\frac{1}{n}-\frac{1}{n'}\right)\alpha v C_p^2\left\{\frac{\alpha+1}{2}v-k_p\right\}\right]$$

$$E_1=\left(\frac{1}{n}-\frac{1}{n'}\right)PC_p^2\left(\alpha v-k_p\right)$$

and

$$f=(n/N).$$

The $\text{MSE}(t_d)$ at (5.3) is minimum for

$$\left.\begin{aligned} w_1&=\left\{(B_1D_1-C_1E_1)/(A_1B_1-C_1^2)\right\}=w_1^*\ \text{(say)}\\ w_2&=\left\{(A_1E_1-C_1D_1)/(A_1B_1-C_1^2)\right\}=w_2^*\ \text{(say)}\end{aligned}\right\} \tag{8.3.29}$$

Thus the resulting minimum MSE of t_d is given by:

$$\mathrm{MSE}_{\min}\left(t_{a7(2)}\right)=\bar{Y}^2\left[1-\frac{\left(B_1D_1^2-2C_1D_1E_1+A_1E_1^2\right)}{\left(A_1B_1-C_1^2\right)}\right] \quad (8.3.30)$$

8.3.6 Estimators Proposed By Haq, Hanif and Ahmad (2011)

Haq et al. (2011) proposed some new estimators for single and two phase sampling.

8.3.6.1 Single Phase sampling estimators

Haq et al. (2011) proposed an estimator for the full information case which is an improved form of the Shabbir and Gupta (2007) estimator, stated as:

$$t_{a9(1)}=d_0\left[\bar{y}-d_1\left(p-p\right)\right]=d_0t^*_{a9(1)} \quad (8.3.31)$$

where $d_0=\dfrac{1}{1+\bar{y}^{-2}MSE\left(t^*_{a9(1)}\right)}$ (Shahbaz and Hanif (2009))

$$t^*_{a9(1)}=\left[\bar{y}-d_1\left(p-p\right)\right] \quad (8.3.32)$$

Using the Shahbaz and Hanif (2009) approach given in (8.3.25) the mean square error of $t_{a9(1)}$ is:

$$MSE\left(t_{a9(1)}\right)\approx\frac{MSE\left(t^*_{a9(1)}\right)}{1+\bar{y}^{-2}MSE\left(t^*_{a9(1)}\right)} \quad (8.3.33)$$

The mean square error of $\left(t^*_{a9(1)}\right)$ is:

$$MSE\left(t^*_{a9(1)}\right)\approx\theta\left[S_y^2+d_1^2S_x^2-2d_1S_yS_xP_{pb}\right] \quad (8.3.34)$$

Optimum value of $d_1=\dfrac{s_yp_{pb}}{s\tau}$

Using optimum value of d_1 in (8.3.34) and simplifying we get:

$$MSE\left(t^*_{a9(1)}\right)\approx\theta\left(1-P_{pb}^2\right)S_y^2 \quad (8.3.35)$$

Using (8.3.35) in (8.3.33) we get:

$$MSE\left(t_{a9(1)}\right) \approx \theta\left(1-P_{pb}^2\right)S_y^2 / 1+\theta\left(1-P_{pb}^2\right)\bar{y}^{-2}S_y^2 \tag{8.3.36}$$

Note that expression (8.3.36) is exact, unlike Shabbir and Gupta (2007).

A regression type estimator by Haq et al. (2011) for two auxiliary attributes is:

$$t_{a10(1)} = \gamma_o\left[\bar{y} - \gamma_1\left(P_1 - p_1\right)\gamma_2\left(P_2 - p_2\right)\right] = \gamma_o t^*_{a10(1)}, \tag{8.3.37}$$

where $\gamma_o = \dfrac{1}{1+\bar{y}^{-2}MSE\left(t^*_{a10(1)}\right)}$ and

$$t_{a10(1)} = \gamma_o\left[\bar{y} - \gamma_1\left(P_1 - p_1\right)\gamma_2\left(P_2 - p_2\right)\right]$$

Using the Shahbaz and Hanif (2009) approach given in (8.3.25) the mean square error is:

$$MSE\left(t_{a10(1)}\right) \approx \frac{MSE\left(t^*_{a10(1)}\right)}{1+\bar{y}^{-2}MSE\left(t^*_{a10(1)}\right)} \tag{8.3.38}$$

The mean square error of $t^*_{a10(1)}$ is:

$$MSE\left(t^*_{a10(1)}\right) \approx \theta\begin{bmatrix} S_y^2 + \gamma_1^2 S_\tau^2 + \gamma_2^2 S_{\tau_2}^2 - 2\gamma_1\gamma_2 S_y S_{\tau_1} P_{pb_1} \\ -2\gamma_2 S_y S_{\tau_2} P_{pb_2} + 2\gamma_1\gamma_2 S_{\tau_1} S_{\tau 2} Q_{12} \end{bmatrix} \tag{8.3.39}$$

Optimum value of γ_1 and γ_2 are:

$$\gamma_1 = \frac{P_1\left(P_{pb_1} - Q_{12}P_{pb_2}\right)S_y}{\left(1-Q_{12}^2\right)S_{\tau_1}} \text{ and } \gamma_2 = \frac{P_2\left(P_{pb_2} - Q_{12}P_{pb_1}\right)S_y}{\left(1-Q_{12}^2\right)S_{\tau_2}}$$

Using the values of γ_1 and γ_2 in (8.3.39) we get:

$$MSE\left(t^*_{a10(1)}\right) \approx \theta\left(1-P_{y.\gamma_1\gamma_2}^2\right)S_y^2 \tag{8.3.40}$$

Using (8.3.40) in (8.3.33) the mean square error will be:

$$MSE\left(t_{a10(1)}\right) \approx \theta\left(1-P_{y.\gamma_1\gamma_2}^2\right)S_y^2 / 1+\theta\left(1-P_{y.\gamma_1\gamma_2}^2\right)S_y^2 \tag{8.3.41}$$

8.3.6.2 Two Phase Sampling Estimators

Haq et al. (2011) proposed some estimators for two phase sampling also for two cases, i.e.:

i) No information case

ii) Partial information case

i) Estimators for No Information Case

Haq et al. (2011) proposed a regression type estimator for two-phase sampling using a single auxiliary attribute, which is an unproved form of the Shabbir and Gupta (2007) estimator, as:

$$t_{a8(2)} = W_o\left[\bar{y}_2 - W_1\left(P_{(2)} - P_{(1)}\right)\right] = W_o t^*_{a8(2)} \quad (8.3.42)$$

where $W_o = \dfrac{1}{1+\bar{y}^{-2}MSE\left(t^*_{a8(2)}\right)}$ and $t^*_{a8(2)} = \left[\bar{y}_2 - W_1\left(P_{(2)} - P_{(1)}\right)\right]$

Using the Shahbaz and Hanif (2009) approach given in (8.3.25), the mean square error is:

$$MSE\left(t_{a8(2)}\right) \approx \frac{MSE\left(t^*_{a8(2)}\right)}{1+\bar{y}^{-2}MSE\left(t^*_{a8(2)}\right)} \quad (8.3.43)$$

or

$$MSE\left(t^*_{a8(2)}\right) \approx \theta_2 S_y^2 + \theta_3\left(W_1 S_\tau^2 - 2W_1 S_y S_\tau P_{pb}\right) \quad (8.3.44)$$

The optimum value of $W_1 = \dfrac{S_y P_{pb}}{S_\tau}$, using the optimum value of W_1 in (8.3.44), the mean square error is:

$$MSE\left(t^*_{a8(2)}\right) \approx \left(Q_2 \_ Q_3 P_{pb}^2\right) S_y^2 \quad (8.3.45)$$

Using (8.3.45) in (8.3.43) the we get:

$$MSE\left(t_{a8(2)}\right) \approx \frac{\left(Q_2 \_ Q_3 P_{pb}^2\right) S_y^2}{1+\left(Q_2 \_ Q_3 P_{pb}^2\right) S_y^2} \quad (8.3.46)$$

(8.3.46) is exact, unlike Shabbir & Gupta (2007).

Another regression type estimator proposed by Haq et al. (2011) for the no information case using two auxiliary attributes is:

$$t_{a9(2)} = \gamma_o^*\left[\bar{y}_2 - \gamma_1\left(P_{1(2)} - P_{1(1)}\right) - \gamma_2\left(P_{2(2)} - P_{2(1)}\right)\right] = \gamma_o^* t^*_{a9(2)}$$

(8.3.47)

where $t^*_{a9(2)} = \left[\bar{y}_2 - \gamma_1\left(P_{1(2)} - P_{1(1)}\right) - \gamma_2\left(P_{2(2)} - P_{2(1)}\right)\right]$

and $\gamma_o^* = \dfrac{1}{1+\bar{y}^{-2}MSE\left(t_{a9(2)}^*\right)}$.

Using the Shahbaz and Hanif (2009) approach given in (8.3.25) the $MSE\left(t_{a9(2)}\right)$ is:

$$MSE\left(t_{a9(2)}\right)\frac{MSE\left(t_{a9(2)}^*\right)}{1+\bar{y}^{-2}MSE\left(t_{a9(2)}^*\right)} \tag{8.3.48}$$

or

$$MSE\left(t_{a9(2)}^*\right) \approx \left\{\begin{matrix} Q_2S_y^2+Q_3\left(\gamma_1^2S_\tau^2+\gamma_2^2S_{\tau 2}^2-2\gamma_1S_yS_{\tau_1}P_{pb_1}\right. \\ \left.-2\gamma_2S_yS_{\tau_2}P_{pb_2}+2\gamma_1\gamma_2S_{\tau_1}S_{\tau 2}Q_{12}\right)\end{matrix}\right\} \tag{8.3.49}$$

The optimum values of γ_1, γ_2 are the same as given in the single phase sampling case for the estimator defined in (8.3.31). Using those optimum values of γ_1 *and* γ_2 in (8.3.49), the $MSE\left(t_{a9(2)}^*\right)$ is:

$$MSE\left(t_{a9(2)}^*\right) \approx \left\{Q_2\left(1-P_{y_o}^2\gamma_1\gamma_2\right)+\theta_1P_{y_o}^2\gamma_1\gamma_2\right\}S_y^2 \tag{8.3.50}$$

Using (8.3.50) in (8.3.48) the mean square error is:

$$MSE\left(t_{a9(2)}\right) \approx \frac{\left\{Q_2\left(1-P_{y_o}^2\gamma_1\gamma_2\right)+\theta_1P_{y_o}^2\gamma_1\gamma_2\right\}S_y^2}{1+\left\{Q_2\left(1-P_{y_o}^2\gamma_1\gamma_2\right)+\theta_1P_{y_o}^2\gamma_1\gamma_2\right\}\bar{y}^{-2}S_y^2} \tag{8.3.51}$$

ii) Estimators for Partial Information Case

Haq et al. (2011) also proposed a regression case estimator for the partial information case as:

$$t_{a10(2)} = \delta_0\left[\bar{y}_2-\delta_1\left(P_{1(2)}-P_1\right)-\delta_2\left(P_{2(2)}-P_{2(1)}\right)\right] = \delta_0\,t_{a10(2)}^* \tag{8.3.52}$$

where $\delta_0 = \dfrac{1}{1+\bar{y}^{-2}MSE\left(t_{a10(2)}^*\right)}$ and

$$t_{a10(2)}^* = \left[\bar{y}_2-\delta_1\left(P_{1(2)}-P_1\right)-\delta_2\left(P_{2(2)}-P_{2(1)}\right)\right]$$

Using the Shahbaz and Hanif (2009) approach in (8.3.25), we get:

$$MSE\left(t_{a10(2)}\right)=\frac{MSE\left(t^*_{a10(2)}\right)}{1-\overline{y}^{-2}MSE\left(t^*_{a10(2)}\right)} \tag{8.3.53}$$

or

$$MSE\left(t^*_{a10(2)}\right)\approx\begin{Bmatrix}\theta_2\left(\overline{y}^2S_y^2+\delta_1^2S_{\tau_1}^2-2\delta_1S_yS_{\tau_1}P_{pb_1}\right)\\+\theta_3\left(\delta_2^2S_{\tau_2}^2-2\delta_2S_yS_{\tau 2}P_{pb_2}+2\delta_1\delta_2S_{\tau_1}S_{\tau 2}Q_{12}\right)\end{Bmatrix} \tag{8.3.54}$$

The optimum values of δ_1 and δ_2 are:

$$\delta_1=\frac{P_1s_y\left(\theta_2P_{pb_1}-\theta_3P_{pb_2}Q_{12}\right)}{s_{\tau_1}\left(\theta_2-\theta_3Q_{12}^2\right.}\text{ and }\delta_2=\frac{\theta_2P_y\left(P_{pb_2}-\theta_3P_{pb_2}Q_{12}\right)}{s_{\tau_1}\left(\theta_2-\theta_3Q_{12}^2\right.}$$

Using the values of δ_1 and δ_2 in (8.3.54), the $MSE\left(t^*_{a10(2)}\right)$ is:

$$MSE\left(t^*_{a10(2)}\right)=\theta_2\left[1-\frac{\theta_2P_{pb_2}+\theta_3P_{pb_2}^2-2\theta_3Q_{12}P_{pb_1}P_{pb_2}}{\left(\theta_2-\theta_3Q_{12}^2\right.}\right]S_y^2 \tag{8.3.55}$$

Using (8.3.55) in (8.3.53), the $MSE\left(t_{a10(2)}\right)$ is:

$$MSE\left(t_{a10(2)}\right)=\frac{\left[\frac{\theta_2\left[1-\theta_2P_{pb_1}^2+\theta_3P_{pb_2}^2-2\theta_3Q_{12}P_{pb_1}P_{pb_2}\right]}{\left(\theta_2-\theta_3Q_{12}^2\right)}\right]S_y^2}{1+\theta_2\left[1-\frac{\theta_2P_{pb_1}^2+\theta_3P_{pb_2}^2-2\theta_3Q_{12}P_{pb_1}P_{pb_2}}{\left(\theta_2-\theta_3Q_{12}^2\right)}\right]\overline{y}^{-2}S_y^2} \tag{8.3.56}$$

Haq et al. (2011) also proposed a family of estimators for the No Information Case as:

$$t_{a11(2)}=g_w\left(\overline{y},v_1,v_{2d}\right) \tag{8.3.57}$$

where $v_1=\frac{P_{1(2)}}{P_1}$, $v_{2d}=\frac{P_{2(2)}}{P_{2(1)}}$ and satisfying certain regularity conditions.

Consider the following estimators of the family defined in (8.3.57):

$$t_{a11(2)}=\overline{y}_2+\alpha_1'\left(v_1-1\right)+\alpha_2'\left(v_{2d}-1\right) \tag{8.3.58}$$

The MSE of $t_{a11(2)}$ is:

$$MSE\left(t_{a11(2)}\right) \approx \theta_2\left[S_y^2 + \alpha_1^{'2}\frac{S_{\tau_1}^2}{P_1^2} + 2\alpha_2^{'}S_y\frac{S\tau_1}{P_1}P_{pb_1}\right]$$
$$+\theta_3\left[\alpha_1^{'2}\frac{S_{\tau_2}^2}{P_2^2} + 2\alpha_2^{'}\overline{y}S_y\frac{S\tau_2}{P_2}P_{pb_2} + 2\alpha_1^{'}\alpha_2^{'}\frac{S_{\tau_1}}{P_1}\frac{S_{\tau_2}}{P_2}Q_{12}\right] \quad (8.3.59)$$

The optimum values of $\alpha_1^{'}$ and $\alpha_2^{'}$ are:

$$\alpha_1^{'} = \frac{-P_1S_y\left(\theta_2P_{pb_1} - Q_{12}Q_{13}P_{pb_2}\right)}{S_{\tau 1}\left(\theta_2 - \theta_3Q_{12}^2\right)} \text{ and } \alpha_2^{'} = \frac{-\theta_2P_2S_y\left(P_{pb_2} - Q_{12}P_{pb_1}\right)}{S_{\tau_2}\left(\theta_2 - \theta_3Q_{12}^2\right)}$$

Using the value of $\alpha_1^{'}$ and $\alpha_2^{'}$ in (8.3.59) the mean square error of $t_{a11(2)}$ is:

$$MSE\left(t_{a11(2)}\right) = \theta_2\left[-1\frac{\theta_2P_{pb_1}^2 + \theta_3P_{pb_2}^2 - 2\theta_3Q_{12}P_{pb_1}P_{pb_2}}{\left(\theta_2 - \theta_3Q_{12}^2\right)}\right]S_y^2 \quad (8.3.60)$$

CHAPTER NINE

FAMILIES OF ESTIMATORS USING MULTIPLE AUXILIARY ATTRIBUTES

9.1 Introduction

Survey statisticians have always been on the verge of developing efficient estimators for estimation of population characteristics. Numerous estimators have been proposed in the literature which utilize information of several auxiliary variables. We have discussed some estimators using information of multiple auxiliary variables in previous chapters of this monograph. Some popular estimators are multiple ratio and regression estimators which have been proved to be more efficient as compared with estimators using one or two auxiliary variables. Ahmad (2008) has given several estimators using multiple auxiliary variables. The ideas developed are a simple extension of classical estimators available in survey sampling. Ahmad (2008) has proposed univariate and multivariate estimators for single and two phase sampling using information of multiple auxiliary variables. He has proposed ratio, regression and product type estimators under various mechanisms. Various authors have proposed estimators using qualitative auxiliary variables. The family of estimator proposed by Jhajj et al. (2006) is a rich family of estimators using information of single qualitative auxiliary variables. The authors have obtained the mean square error of the proposed family and have listed various members of the proposed family. In previous chapter we discussed various families of estimators for single and two phase sampling which use information of one and two qualitative auxiliary variables. The families of estimators discussed there have natural extension to the case of several qualitative auxiliary variables. We now discuss some of the estimators which use information on several qualitative variables. We will also present some families of estimators following the lines of Jhajj et al. (2006) and Hanif et al. (2009) for single phase as well as for two phase sampling.

9.2 Estimators for Single Phase Sampling

Various estimators have been proposed from time to time which use information of several auxiliary variables and that can be used for single phase sampling. The famous generalized regression estimator by Cassel et al. (1977) has attracted various survey statisticians in developing new estimators for single phase sampling using multiple auxiliary variables. Situations do arise when information of qualitative auxiliary variables is available and survey estimates have to be obtained using that available information. The estimators which use information of quantitative auxiliary variables are not helpful in that sort of situation, and we have to look for alternative estimation methodologies. The family of estimators proposed by Jhajj et al. (2006) is perhaps treated as the first step in developing estimators using information of qualitative auxiliary variables. Hanif et al. (2009) extended the idea of Jhajj et al. (2006) and they proposed a family of estimators using information of two auxiliary variables. In the following we present some of the recent estimators for single phase sampling proposed by Hanif et al. (2009) and Hanif et al. (2010). The family of estimators proposed by Hanif et al. (2009) using information of several qualitative auxiliary variables will also be discussed.

9.2.1 Hanif, Haq and Shahbaz (2009, 2010) Estimators Using Multi-Auxiliary Attributes

Consider that information on k qualitative auxiliary variables is available and we need to estimate population mean or total. Following the idea of multiple regression estimators with several quantitative variables, Hanif et al. (2009, 2010) have proposed different estimators by using information of several qualitative auxiliary variables.

9.2.1.1 Regression Type Estimator

The first estimator proposed by Hanif et al. (2009) is a regression type estimator but it is based upon qualitative auxiliary variables. The estimator is given as:

$$t_{a1(1)} = \bar{y} + \sum_{j=1}^{k} \alpha_j \left(v_j - 1\right) \tag{9.2.1}$$

where $v_j = p_j / P_j$. The above estimator can also be written as:

$$t_{a1(1)} = \overline{y} + \underline{\alpha}'(\underline{v} - \underline{1}) = \overline{y} + \underline{\alpha}'\underline{\varphi}, \tag{9.2.2}$$

where $\boldsymbol{\alpha}$ and $\mathbf{v}$ are (k x 1) vectors and $\boldsymbol{\varphi} = \mathbf{v} - \mathbf{1}$. The mean square error can be readily put in the form:

$$MSE(t_{a1(1)}) = E\left[\overline{e}_y^2 + \underline{\alpha}'\underline{\varphi}\underline{\varphi}'\underline{\alpha} + 2\underline{\alpha}'\underline{\varphi}\overline{e}_y\right].$$

Applying expectation and using the fact that $E(\varphi\varphi') = \boldsymbol{\Phi}_\tau$ and $E(\varphi\overline{e}_y) = \phi_{y\tau}$, the mean square error can be written as:

$$MSE(t_{a1(1)}) \approx \theta\left[\overline{Y}^2C_y^2 + \underline{\alpha}'\boldsymbol{\Phi}_{\boldsymbol{\tau}}\underline{\alpha} + 2\underline{\alpha}'\underline{\phi}_{y\tau}\right].$$

The value of $\boldsymbol{\alpha}$ which minimizes the above mean square error can be easily obtained and the value turns out to be $\alpha = -\boldsymbol{\Phi}_\tau^{-1}\phi_{y\tau}$. Using this value the mean square error of (9.2.1) turns out to be:

$$MSE(t_{a1(1)}) = \theta\overline{Y}^2C_y^2\left(1 - \rho^2_{y.\tau_1\tau_2\ldots\tau_k}\right), \tag{9.2.3}$$

where $\rho^2_{y.\tau_1\tau_2\ldots\tau_k}$ is the squared multiple point bi–serial correlation between Y and combined effect of all the auxiliary attributes. The mean square error given in (9.2.3) is obviously smaller than the mean square error of any member of the family of estimators proposed by Jhajj et al. (2006) and by Hanif et al. (2009).

Ratio and product estimators are popular estimation methods which are based upon available information of some auxiliary variables. Hanif et al. (2010) have demonstrated the use of multiple auxiliary attributes in ratio and product estimators.

9.2.1.1 Ratio Type Estimator

The following ratio estimator using information of multiple qualitative auxiliary variables has been proposed by Hanif et al. (2010):

$$t_{a12(1)} = \overline{y}\left(\frac{P_1}{p_1}\right)\left(\frac{P_2}{p_2}\right)\cdots\cdots\left(\frac{P_k}{p_k}\right) = \overline{y}\prod_{j=1}^{k}\left(P_j/p_j\right). \tag{9.2.4}$$

Hanif et al. (2010) has shown that the mean square error of (8.2.4) can be readily obtained by expanding (9.2.4) as:

$$MSE\left(t_{a12(1)}\right) \approx E\left(\bar{e}_y - \sum_{j=1}^{k} \frac{\bar{Y}}{P_j} \bar{e}_{\tau_j}\right)^2.$$

Opening the square and applying expectation, the mean square error of (9.2.4) is:

$$MSE\left(t_{a12(1)}\right) \approx \theta \bar{Y}^2 \left[C_y^2 + \sum_{j=1}^{k} C_{\tau_j}^2 - 2\sum_{j=1}^{k} C_y C_{\tau_j} \rho_{Pb_j} + 2\sum_{j \neq \psi = 1}^{k} C_{\tau_j} C_{\tau_\psi} Q_{j\psi}\right]$$

(9.2.5)

The product estimator parallel to (9.2.4) has also been proposed by Hanif et al. (2010) and is given as:

$$t_{a13(1)} = \bar{y}\left(\frac{p_1}{P_1}\right)\left(\frac{p_2}{P_2}\right) \cdots\cdots \left(\frac{p_k}{P_k}\right) = \bar{y}\prod_{j=1}^{k}\left(p_j / P_j\right). \qquad (9.2.6)$$

Proceeding in the above manner, Hanif et al. (2010) has shown that the mean square error of (9.2.6) is:

$$MSE\left(t_{a13(1)}\right) \approx \theta \bar{Y}^2 \left[C_y^2 + \sum_{j=1}^{k} C_{\tau_j}^2 + 2\sum_{j=1}^{k} C_y C_{\tau_j} \rho_{Pb_j} + 2\sum_{j \neq \psi = 1}^{k} C_{\tau_j} C_{\tau_\psi} Q_{j\psi}\right].$$

(9.2.7)

The same argument comes into play for choosing between (9.2.4) and (9.2.6) which is given for choosing between classical ratio and product estimators. From the comparison of (9.2.3), (9.2.5) and (9.2.7) we can readily see that (9.2.1) has the least mean square error among the three proposed estimators by Hanif et al. (2010). Hanif et al. (2010) has also proposed a family of estimators for single phase sampling. We discuss the family in the following.

9.2.2 Family of Estimators by Hanif, Haq and Shahbaz (2009) for Single Phase Sampling

A general family of estimators has been proposed by Hanif et al. (2009) by extending the argument of Jhajj et al. (2006). The family of estimators has been proposed for the estimation of population mean using information of multiple qualitative auxiliary variables. The proposed family has the general set up of:

$$t_{a14(1)} = g_{\omega}\left(\overline{y}, v_1, v_2, \ldots, v_k\right), \tag{9.2.8}$$

where $v_j = p_j / P_j ; v_j > 0$ and $g_{\omega}\left(\overline{y}, v_1, v_2, \ldots, v_k\right)$ is any parametric function of available variables such that $g_{\omega}\left(\overline{y}, v_1, v_2, \ldots, v_k\right)$. Some possible members of the family (8.2.8) suggested by Hanif et al. (2009) are:

i) $$g_{\omega}\left(\overline{y}, v_{1,} v_2, \ldots, v_k\right) = \overline{y} + \sum_{j=1}^{k} \alpha_j \left(v_j - 1\right), \tag{9.2.9}$$

ii) $$g_{\omega}\left(\overline{y}, v_{1,} v_2, \ldots, v_k\right) = \overline{y} \exp\left\{\sum_{j=1}^{k} \alpha_j \left(v_j - 1\right)\right\}, \tag{9.2.10}$$

and

iii) $$g_{\omega}\left(\overline{y}, v_{1,} v_2, \ldots, v_k\right) = \overline{y} \prod_{j=1}^{k} v_j^{\alpha_j} . \tag{9.2.11}$$

The mean square error of (9.2.8) can be obtained by using the argument of Jhajj et al. (2006) and expanding (9.2.8) by using second order Taylor expansion series as:

$$\begin{aligned} t_{a14(1)} = \overline{Y} + \left(\overline{y} - \overline{Y}\right) g_{0a} + \sum_{j=1}^{k} \left(v_j - 1\right) g_{ja} + \frac{1}{2} \Bigg\{ \left(\overline{y} - \overline{Y}\right)^2 g_{00a} \\ + \sum_{j=1}^{k} \left(v_j - 1\right)^2 g_{jja} + 2 \sum_{j=1}^{k} \left(\overline{y} - \overline{Y}\right)\left(v_j - 1\right) g_{0ja} \\ + 2 \sum\sum_{j<h=1}^{k} \left(v_j - 1\right)\left(v_h - 1\right) g_{jha} \Bigg\} + \cdots, \end{aligned}$$

where $g_{ja}; j = 0,1,2,\ldots k;$ are first derivatives of $t_{a14(1)}$ w.r.t. $g_{\omega}\left(\overline{y}, v_1, v_2, \ldots, v_k\right)$ and $g_{jha}; \left(j,h\right) = 0,1,2,\ldots,k;$ are corresponding second order derivatives. Applying the expectation we can see that $E\left(t_{14(1)}\right) = \overline{Y}$. Hanif et al. (2009) has further shown that the mean square error of (9.2.8) to $O\left(1/n\right)$ is given as:

$$MSE\left(t_{a14(1)}\right) = \theta \overline{Y}^2 C_y^2 \left(1 - \rho^2_{y.\tau_1\tau_2\ldots\tau_k}\right). \tag{9.2.12}$$

The mean square error given in (9.2.12) is the same as the mean square error given in (9.2.3). Hanif et al. (2009) has commented that any member of the family will have the same mean square error given in (9.2.12). The argument of Hanif et al. (2009) can be justified from (9.2.3). The estimate of the mean square error of (8.2.8) is readily written as:

$$MSE\left(t_{a14(1)}\right)=\theta\overline{y}^2c_y^2\left(1-r^2_{y.\tau_1\tau_2...\tau_k}\right). \tag{9.2.13}$$

Any member of the family $t_{14(1)}$ alongside (9.2.13) can be used to construct a confidence interval for population mean.

9.3 Estimators for Two-Phase Sampling

The idea behind the use of multiple qualitative auxiliary variables in single phase sampling can be easily extended to two-phase sampling. A simple demonstration of use of a single qualitative auxiliary variable in two-phase sampling has been given by Jhajj et al. (2006). We have presented that idea along with its extension to two qualitative auxiliary variables in the previous chapter. We now present some of the estimators for estimation of population mean by using information of multiple qualitative auxiliary variables.

9.3.1 Hanif-Haq- Shahbaz (2010) Estimators

We have presented two estimators proposed by Hanif et al. (2010) which use information of multiple qualitative auxiliary variables for single phase sampling in section 8.2.1. The idea of the presented estimators can be easily extended to two phase sampling under various mechanisms discussed in 8.3.4. Some of the proposed estimators by Hanif et al. (2010) are discussed in the following.

9.3.1.1 Estimators for No Information Case

When information of all auxiliary variables is available only for the second phase, then this situation is peculiar for two-phase sampling. Hanif et al. (2009, 2010) proposed the following estimators for two-phase sampling when information of auxiliary attributes is available for second-phase only. The authors have proposed ratio, product and regression estimators using information on several qualitative auxiliary variables.

(a) Ratio Estimator for No information Case

The ratio estimator proposed by Hanif et al. (2010), when information on auxiliary attributes is available for second phase only, is given as:

$$t_{a12(2)} = \bar{y}_2 \left(\frac{p_{1(1)}}{p_{1(2)}}\right)\left(\frac{p_{2(1)}}{p_{2(2)}}\right)\cdots\cdots\left(\frac{p_{k(1)}}{p_{k(2)}}\right) = \bar{y}_2 \prod_{j=1}^{k}\left(p_{j(1)}/p_{j(2)}\right). \quad (9.3.1)$$

The mean square error of (9.3.1) is given as:

$$MSE\left(t_{a12(2)}\right) \approx E\left(\bar{e}_{y_2} - \sum_{j=1}^{k}\frac{\bar{Y}}{P_j}\left(\bar{e}_{\tau_{j(2)}} - \bar{e}_{\tau_{j(1)}}\right)\right)^2.$$

Opening the square and applying expectations, Hanif et al. (2010) have shown that the mean square error of (9.3.1) is given as:

$$MSE\left(t_{a12(2)}\right) \approx \bar{Y}^2\left[\theta_2 C_y^2 + \theta_3\left\{\sum_{j=1}^{k} C_{\tau_j}^2 - 2\sum_{j=1}^{k} C_y C_{\tau_j}\rho_{Pb_j} + 2\sum_{j\neq\psi=1}^{k} C_{\tau_j} C_{\tau_\psi} Q_{j\psi}\right\}\right] \quad (9.3.2)$$

The estimate of (9.3.2) can be easily written by replacing population parameters with sample estimates.

(b) Product- Ratio Estimator for No Information Case

The product estimator proposed by Hanif et al. (2010) is given as:

$$t_{a13(2)} = \bar{y}_2 \left(\frac{p_{1(2)}}{p_{1(1)}}\right)\left(\frac{p_{2(2)}}{p_{2(1)}}\right)\cdots\cdots\left(\frac{p_{k(2)}}{p_{k(1)}}\right) = \bar{y}_2 \prod_{j=1}^{k}\left(p_{j(2)}/p_{j(1)}\right), \quad (9.3.3)$$

The mean square error of (9.3.3) is readily written as:

$$MSE\left(t_{a13(2)}\right) \approx \bar{Y}^2\left[\theta_2 C_y^2 + \theta_3\left\{\sum_{j=1}^{k} C_{\tau_j}^2 + 2\sum_{j=1}^{k} C_y C_{\tau_j}\rho_{Pb_j} + 2\sum_{j\neq\psi=1}^{k} C_{\tau_j} C_{\tau_\psi} Q_{j\psi}\right\}\right].$$

(c) Regression Estimator for No Information Case

The regression estimator for the no information case proposed by Hanif et al. (2009) is given as:

$$t_{a14(2)} = \bar{y}_2 + \sum_{j=1}^{k}\alpha_j\left(v_{jd} - 1\right), \quad (9.3.4)$$

where $v_{jd} = p_{j(1)}/p_{j(2)}$. Following the earlier arguments, Hanif et al. (2009) have shown that the mean square error of (9.3.4) is:

$$MSE\left(t_{a14(2)}\right) \approx \bar{Y}^2 C_y^2\left\{\theta_2\left(1 - \rho_{y.\tau_1\tau_2\ldots\tau_k}^2\right) + \theta_1\rho_{y.\tau_1\tau_2\ldots\tau_k}^2\right\}. \quad (9.3.5)$$

The estimate of (9.3.5) can be immediately written by replacing the population parameters with their sample counterparts.

9.3.1.2 Estimator for Partial Information Case

When some information from the population and sample information from the sample either at first phase or second phase is available, then Ahmad et al. (2010) suggested the following various estimators.

(a) Ratio Estimator for Partial Information Case-I

The ratio estimator for the partial information case by Hanif et al. (2010) has the form:

$$t_{a15(2)} = \bar{y}_2\left(\frac{P_1}{p_{1(1)}}\right)\left(\frac{P_2}{p_{2(1)}}\right)\cdots\cdots\left(\frac{P_m}{p_{m(1)}}\right)\left(\frac{p_{(m+1)(1)}}{p_{(m+1)(2)}}\right)\cdots\cdots\left(\frac{p_{k(1)}}{p_{k(2)}}\right),$$

or $$t_{a15(2)} = \bar{y}_2\prod_{j=1}^{m}\left(P_j/p_{j(1)}\right)\prod_{h=m+1}^{k}\left(p_{h(1)}/p_{h(2)}\right). \qquad (9.3.6)$$

The mean square error of the above estimator can be expanded as:

$$MSE\left(t_{a15(2)}\right) = E\left(\bar{e}_{y_2} - \sum_{j=1}^{m}\frac{\bar{Y}}{P_j}\bar{e}_{\tau_{j(1)}} - \sum_{j=m+1}^{k}\frac{\bar{Y}}{P_j}\left(\bar{e}_{\tau_{j(2)}} - \bar{e}_{\tau_{j(1)}}\right)\right)^2.$$

Using the notations from section 1.4, the mean square error of (9.3.6) is:

$$MSE\left(t_{a15(2)}\right) \approx \bar{Y}^2\left[\theta_2\left\{C_y^2 + \sum_{j=m+1}^{k}C_{\tau_j}^2 - 2\sum_{j=m+1}^{k}C_yC_{\tau_j}\rho_{Pb_j} + 2\sum_{j\neq\psi=m+1}^{k}C_{\tau_j}C_{\tau_\psi}Q_{j\psi}\right\}\right.$$
$$\theta_1\left\{\left(\sum_{j=1}^{m}C_{\tau_j}^2 - \sum_{j=m+1}^{k}C_{\tau_j}^2\right) - 2\left(\sum_{j=1}^{m}C_yC_{\tau_j}\rho_{Pb_j} - \sum_{j=m+1}^{k}C_yC_{\tau_j}\rho_{Pb_j}\right)\right.$$
$$\left.\left. +2\left(\sum_{j\neq\psi=1}^{m}C_{\tau_j}C_{\tau_\psi}Q_{j\psi} - \sum_{j\neq\psi=m+1}^{k}C_{\tau_j}C_{\tau_\psi}Q_{j\psi}\right)\right\}\right].$$

(9.3.7)

(b) Ratio Estimator for Partial Information Case-II

Another estimator proposed by Hanif et al. (2010) for the partial information case is given as:

$$t_{a16(2)} = \bar{y}_2\left(\frac{P_1}{p_{1(2)}}\right)\left(\frac{P_2}{p_{2(2)}}\right)\cdots\cdots\left(\frac{P_m}{p_{m(2)}}\right)\left(\frac{p_{(m+1)(1)}}{p_{(m+1)(2)}}\right)\cdots\cdots\left(\frac{p_{k(1)}}{p_{k(2)}}\right),$$

or $t_{a16(2)} = \bar{y}_2 \prod_{j=1}^{m} \left(P_j / p_{j(2)}\right) \prod_{h=m+1}^{k} \left(p_{h(1)} / p_{h(2)}\right).$ (9.3.8)

The mean square error of (9.3.9) is:

$$MSE\left(t_{a16(2)}\right) \approx \bar{Y}^2 \left[\theta_2 \left\{C_y^2 + \sum_{j=1}^{m} C_{\tau_j}^2 - 2\sum_{j=1}^{m} C_y C_{\tau_j} \rho_{Pb_j} + 2 \sum_{j \neq \psi = 1}^{m} C_{\tau_j} C_{\tau_\psi} Q_{j\psi} \right\} \right.$$
$$+\theta_3 \left\{ \sum_{j=m+1}^{k} C_{\tau_j}^2 - 2 \sum_{j=m+1}^{k} C_y C_{\tau_j} \rho_{Pb_j} + 2 \sum_{\substack{j=m+1 \\ \psi \geq m+1}} C_{\tau_j} C_{\tau_\psi} Q_{j\psi} \right.$$
$$\left. \left. 2 \sum_{j \neq \psi = m+1}^{k} C_{\tau_j} C_{\tau_\psi} Q_{j\psi} \right\} \right].$$

(9.3.9)

(c) Product- Ratio Estimator for Partial Information Case-III

Hanif et al. (2010) have also proposed the product estimators in two-phase sampling for the partial information case. The estimator goes parallel to the ratio estimator and is given below:

$$t_{a17(2)} = \bar{y}_2 \left(\frac{p_{1(2)}}{P_1}\right)\left(\frac{p_{2(2)}}{P_2}\right) \cdots\cdots \left(\frac{p_{m(2)}}{P_m}\right)\left(\frac{p_{(m+1)(2)}}{P_{(m+1)(1)}}\right) \cdots\cdots \left(\frac{p_{k(2)}}{P_{k(1)}}\right),$$

or

$$t_{a17(2)} = \bar{y}_2 \prod_{j=1}^{m} \left(p_{j(2)} / P_j\right) \prod_{h=m+1}^{k} \left(p_{h(2)} / p_{h(1)}\right). \quad (9.3.10)$$

The mean square error of (9.3.11), obtained by Hanif et al. (2010), is:

$$MSE\left(t_{a17(2)}\right) \approx \bar{Y}^2 \left[\theta_2 \left\{C_y^2 + \sum_{j=m+1}^{k} C_{\tau_j}^2 + 2 \sum_{j=m+1}^{k} C_y C_{\tau_j} \rho_{Pb_j} + 2 \sum_{j \neq \psi = m+1}^{k} C_{\tau_j} C_{\tau_\psi} Q_{j\psi} \right\} \right.$$
$$\theta_1 \left\{ \left(\sum_{j=1}^{m} C_{\tau_j}^2 + \sum_{j=m+1}^{k} C_{\tau_j}^2 \right) + 2\left(\sum_{j=1}^{m} C_y C_{\tau_j} \rho_{Pb_j} + \sum_{j=m+1}^{k} C_y C_{\tau_j} \rho_{Pb_j} \right) \right.$$
$$\left. \left. +2\left(\sum_{j \neq \psi = 1}^{m} C_{\tau_j} C_{\tau_\psi} Q_{j\psi} + \sum_{j \neq \psi = m+1}^{k} C_{\tau_j} C_{\tau_\psi} Q_{j\psi} \right) \right\} \right].$$

(9.3.11)

The estimate of (9.3.7), (9.3.9) and (9.3.11) may be readily written. The regression method of estimation has its significance in two-phase sampling.

(d) Regression Estimator for Partial Information Case-IV

The following regression estimator in two phase sampling has been proposed by Hanif et al. (2009) for the partial information case:

$$t_{a18(2)} = \bar{y}_2 + \sum_{j=1}^{m} \alpha_j \left(v_j - 1\right) + \sum_{j=m+1}^{k} \beta_j \left(w_j - 1\right), \tag{9.3.12}$$

where $v_j = p_{j(1)} / P_j$ and $w_j = p_{j(2)} / p_{j(1)}$. The estimator (9.3.12) may be written as:

$$t_{a18(2)} = \bar{y}_2 + \boldsymbol{\alpha}' \left(\mathbf{v_j} - \mathbf{1}\right) + \boldsymbol{\beta}' \left(\mathbf{w_j} - \mathbf{1}\right) = \bar{y}_2 + \boldsymbol{\alpha}' \boldsymbol{\varphi}_1 + \boldsymbol{\beta}' \boldsymbol{\varphi}_2 .$$

Using the notations from section 1.4, we can immediately see that the mean square error of (9.3.12) is given as:

$$MSE\left(t_{a18(2)}\right) = E\left[\bar{e}_{y_2}^2 + \alpha' \varphi_1 \varphi_1' \alpha + \beta' \varphi_2 \varphi_2' \beta + 2\alpha' \varphi_1 \bar{e}_{y_2} + 2\beta' \varphi_2 \bar{e}_{y_2} + \alpha' \varphi_1 \varphi_2' \beta + \beta' \varphi_2 \varphi_1' \alpha\right]$$

Applying expectation and after simplification we have:

$$MSE\left(t_{a18(2)}\right) \approx \left[\theta_2 \bar{Y}^2 C_y^2 + \theta_1 \left\{\boldsymbol{\alpha}' \boldsymbol{\Phi}_{\boldsymbol{\tau}_1} \boldsymbol{\alpha} + 2\boldsymbol{\alpha}' \phi_{\mathbf{y}\boldsymbol{\tau}_1}\right\} + \left(\theta_2 - \theta_1\right) \left\{\boldsymbol{\beta}' \boldsymbol{\Phi}_{\boldsymbol{\tau}_2} \boldsymbol{\beta} + 2\underline{\boldsymbol{\alpha}}' \phi_{\mathbf{y}\boldsymbol{\tau}_2}\right\}\right] \tag{9.3.13}$$

The optimum values of $\boldsymbol{\alpha}$ and $\boldsymbol{\beta}$ which minimize the above mean square error are $\boldsymbol{\alpha} = -\boldsymbol{\Phi}_{\tau_1}^{-1} \phi_{y\tau_1}$ and $\boldsymbol{\beta} = -\boldsymbol{\Phi}_{\tau_2}^{-1} \phi_{y\tau_2}$. Using these values in (9.3.13) and simplifying, the mean square error will be:

$$MSE\left(t_{a18(2)}\right) = \bar{Y}^2 C_y^2 \left[\theta_2 \left\{1 - \rho^2_{y.\tau_{m+1}\tau_{m+2}\cdots\tau_k}\right\} + \theta_1 \left\{\rho^2_{y.\tau_{m+1}\tau_{m+2}\cdots\tau_k} - \rho^2_{y.\tau_1\tau_2\cdots\tau_m}\right\}\right] \tag{9.3.14}$$

9.3.2 Families of Estimators by Hanif, Haq and Shahbaz (2009)

We have presented some estimators of population mean which use information of multiple qualitative auxiliary variables. We have seen that the regression estimator has a smaller mean square error as compared with ratio and product estimators both in single phase and two phase sampling. The presented estimators for two phase sampling can be viewed as the members of a more general family of estimators, as discussed in section

9.2.2. We now present another family of estimators for two-phase sampling using information of several qualitative auxiliary variables.

9.3.2.1 Family of Estimators for No Information Case

The no information case in two phase sampling has been already described as the case when information of no auxiliary variable is available for complete population and we have to rely only on information of first phase and second phase sample. We have presented some particular estimators for the no information case in two phase sampling in section 9.3.1. Hanif et al. (2009) have proposed a family of estimators for two phase sampling when information of auxiliary variables is available for first and second phase sample only. The proposed family of estimators is given as:

$$t_{a19(2)} = g_{\omega}\left(\bar{y}_2, v_{1d}, v_{2d}, \ldots, v_{kd}\right), \tag{9.3.15}$$

where $v_{jd} = p_{j(2)} / p_{j(1)}$ and function $g_{\omega}\left(\bar{y}_2, v_{1d}, v_{2d}, \ldots, v_{kd}\right)$ is such that $g_{\omega}\left(\bar{Y}, 1, 1, \ldots, 1\right) = \bar{Y}$. Hanif et al. (2009) have presented the following members of the family (9.3.15):

i) $$g_{\omega}\left(\bar{y}_2, v_{1d}, v_{2d} \ldots, v_{kd}\right) = \bar{y}_2 + \sum_{j=1}^{k} \alpha_j \left(v_{jd} - 1\right) \tag{9.3.16}$$

ii) $$g_{\omega}\left(\bar{y}_2, v_{1d}, v_{2d} \ldots, v_{kd}\right) = \bar{y}_2 \exp\left\{\sum_{j=1}^{k} \alpha_j \left(v_{jd} - 1\right)\right\} \tag{9.3.17}$$

and

iii) $$g_{\omega}\left(\bar{y}_2, v_{1d}, v_{2d} \ldots, v_k\right) = \bar{y}_2 \prod_{j=1}^{k} v_{jd}^{\alpha_j} . \tag{9.3.18}$$

Hanif et al. (2009) have shown that the mean square error of (9.3.16) to $O(1/n)$ is:

$$MSE\left(t_{a19(2)}\right) = \bar{Y}^2 C_y^2 \left\{\theta_2 \left(1 - \rho_{y.\tau_1\tau_2\ldots\tau_k}^2\right) + \theta_1 \rho_{y.\tau_1\tau_2\ldots\tau_k}^2\right\}. \tag{9.3.19}$$

Hanif et al. (2009) have further argued that any member of (9.3.15) will have the same mean square error as given in (9.3.19).

9.3.2.2 Family of Estimators for Partail Information Case

We have already presented the situation of the partial information case as the situation when information on some of the auxiliary variables is available for the complete population and information of some of the auxiliary variables is only available for the first phase and second phase sample. Hanif et al. (2009) has presented a family of estimators for two phase sampling using multiple auxiliary attributes. The proposed family goes along the lines of the family of estimators proposed by Jhajj et al. (2006) and is given as:

$$t_{a20(2)} = g_{\omega}\left(\bar{y}_2, v_1, v_2, \ldots, v_m, w_{m+1}, w_{m+2}, \ldots, w_k\right), \tag{9.3.20}$$

where $v_j = p_{j(1)}/P_j$ and $w_j = p_{j(2)}/p_{j(1)}$. The function (9.3.20) possesses the same properties as we have discussed in section 8.2.2. Some of the possible members of the family of estimators given in (9.3.20) are:

i) $$g_{\omega}\left(\bar{y}_2, v_1, \ldots, v_m, w_{m+1}, \ldots, w_k\right) = \bar{y}_2 + \sum_{j=1}^{m}\alpha_j\left(v_j - 1\right) + \sum_{j=m+1}^{m}\beta_j\left(w_j - 1\right) \tag{9.3.21}$$

ii) $$g_{\omega}\left(\bar{y}_2, v_1, \ldots, v_m, w_{m+1}, \ldots, w_k\right) = \bar{y}_2 \exp\left\{\sum_{j=1}^{m}\alpha_j\left(v_j - 1\right) + \sum_{j=m+1}^{m}\beta_j\left(w_j - 1\right)\right\}, \tag{9.3.22}$$

and

iii) $$g_{\omega}\left(\bar{y}_2, v_1, \ldots, v_m, w_{m+1}, \ldots, w_k\right) = \bar{y}_2 \prod_{j=1}^{m} v_j^{\alpha_j} \prod_{j=m+1}^{k} w_j^{\beta_j}, \tag{9.3.23}$$

Hanif et al. (2009) have shown that the mean square error of (9.3.20) can be obtained by expanding it using second order Taylor expansion series as in section 9.2.2. Hanif et al. (2009) have further obtained the following mean square error of (9.3.20):

$$MSE\left(t_{a20(2)}\right) = \bar{Y}^2 C_y^2\left[\theta_2\left\{1 - \rho^2_{y.\tau_{m+1}\tau_{m+2}\ldots\tau_k}\right\} + \theta_1\left\{\rho^2_{y.\tau_{m+1}\tau_{m+2}\ldots\tau_k} - \rho^2_{y.\tau_1\tau_2\ldots\tau_m}\right\}\right]. \tag{9.3.24}$$

Hanif et al. (2009) have argued that the mean square error of any member of the family (9.3.20) will be the same and equal to (9.3.24). The argument of the authors can be verified by looking at estimator (9.3.15), which is presented as a member of family (9.3.20) and is given in (9.3.21). The mean square error of (9.3.13) is given in (9.3.20), which is exactly the same as the mean square error given in (9.3.24). Hanif et al. (2009) have further commented that estimate of (9.3.24) is readily obtained by replacing population parameters with sample estimates in (9.3.24). The confidence interval for the population mean is then readily available by using any member of (9.3.20) and estimate of (9.3.24).

REFERENCES

Abd-Elfattah, A. M., El-Sherpieny, E.A., Mohamed, S.M. and Abdou, O.F. (2010). Improvement in estimating the population mean in simple random sampling using information on auxiliary attribute, *Appl. Math. Comput*. DOI:10.1016/j.amc.2009.12.041.

Adhvaryu, D., Gupta, P.C. (1983). On some alternative sampling strategies using auxiliary information. *Metrika*, 30, 4, 217-226.

Agarwal, S.K. and Kumar, S. (1980). Use of multivariate auxiliary information in selection of units in probability proportional to size with replacement. *J. Indian Soc. Agri. Statist.,* 32, 81-86.

Ahmad, Z. (2008). *Generalized Multivariate Ratio and Regression Estimators for Multi-Phase Sampling.* Unpublished Ph.D. thesis, National College of Business Administration and Economics, Lahore.

Ahmad, Z. and Hanif, M. (2010). Generalized Multi-Phase Multivariate Regression Estimators for Partial Information Case using Multi-Auxiliary Variables. Accepted. *World Applied Sciences Journal,* 10(3), 370-379.

Ahmad, Z., Hanif, M. and Ahmad, M. (2007). Some Generalized Chain Ratio Estimators for Two-Phase Sampling using Multi-Auxiliary Variables. Submitted for publication.

Ahmad, Z., Hanif, M. and Ahmad, M. (2009). Generalized Regression-Cum-Ratio Estimators For Two-Phase Sampling Using Multi-Auxiliary Variables. *Pak. J. Statist.,* 25(2), 93-106.

Ahmad, Z., Hanif, M. and Ahmad, M. (2010). Generalized Multi-Phase Multivariate Ratio Estimators for Partial Information Case using Multi-Auxiliary Variables. *Commun. of the Kor. Statist. Soc.,* 17(5), 625-637.

Ahmad, Z., Hanif, M. and Ahmad, M. (2010). Generalized Regression-in Regression Estimators for Two-Phase Sampling using Multi-Auxiliary Variables. *J. Statist. Theo. and Applica.,* 9(3), 443-457.

Ahmad, Z., Hussin, A.G. and Hanif, M. (2010). Generalized Multivariate Regression Estimators for Multi-Phase Sampling using Multi-Auxiliary Variables. *Pak. J. Statist.,* 26(4), 569-583.

Ahmad, Z., Saqib, M.A., Hussin, A.G. and Hanif, M. (2010). Estimators for Two-Phase Sampling using Multi-Auxiliary Variables. Submitted in *Pak. J. Statist.*

Arora, S. and Bansi, L. (1989). *New Mathematical Statistics*. Satya Prakashan, New Delhi.

Bahl, S. and Tuteja, R.K. (1991): Ratio and Product type exponential estimator. *Information and Optimization Sciences*, 7(1), 159-163.

Bano, S. (2009). *An empirical comparison of some well known estimators in two phase sampling*. A thesis submitted to Allama Iqbal Open University, Islamabad.

Beale, E.M.L. (1962). Some uses of computers in operations research. *Industrielle Organisation,* 31, 51-52.

Bedi, P.K. (1985). On two-phase multivariate sampling estimator. *J. Ind. Soc. Agri. Statist.*, 37(2), 158-162.

Bowley, A.L. (1913). Working class households in Reading. *J. Roy. Statist. Soc.,* 76, 672-701.

—. (1926). Measurements of precision attained in sampling. *Bull. Inst. Inte. Statist.*, 22, 1-62.

Breidt J. and Fuller, W.A. (1993). Regression weighting for multiplicative samples. *Sankhy, 6,* 55, 297-309.

Brewer, K.R.W. and Hanif, M. (1983). *Sampling with unequal probabilities*. Springer Verlag, New York.

Brewer, K.R.W. (1963). Ratio estimation and finite populations: some results deducible from the assumption of an underlying stochastic process. *Austral. J. Statist.*, 5, 93-105.

Bru, B. (1988). Estimations Laplaciennes. Un exemple: la recherche de la population d'un grand Empire, 1785–1812, in *Estimation et sondages. Cinq contributions à l'histoire de la statistique*, ed. Mairesse, J., Paris: Economica, pp. 7–46.

Cassel, C.M., Sarndal, C.E. and Wretman, J.H. (1977). *Foundations of inference in survey sampling*. Wiley-Interscience.

Chand, L. (1975). *Some ratio type estimators based on two or more auxiliary variables*. Unpublished Ph.D. Thesis, Iowa State University, Ames, Iowa (USA).

Chaubey, Y.P., Singh, M., Dwivedi, T.D. (1984). A note on and optimality property of the regression estimator. *Biometrical Journal,* 26(4), 465.

Chaudhuri, A. and Roy, D. (1994). Model assisted survey sampling strategy in two phases. *Metrika*, 41, 355-362.

Cochran, W.G. (1940). The estimation of the yields of cereal experiments by sampling for the ratio of grain to total produce. *J. Agricultural Sc.*, 30, 262-275.

—. (1942). Sampling theory when the sampling units are of unequal sizes. *J. Amer. Statist. Assoc.,* 37, 199-212.

—. (1977). *Sampling Techniques.* J. Wiley New York.

—. (1977). *Sampling techniques.* John Wiley and Sons New York.

Dalenius, T. (1950). The problem of optimum stratification. *Skandinavisk Aktuartidskrijt,* 33,203-213.

—. (1954). Multivariate sampling problem. *Skandinavisk Aktuartidskrijt,* 36, 92-122.

Das, A.K and Tripathi, T.P. (1979). *A class of sampling strategies for population mean using knowledge on the variance of an auxiliary character.* Stat-Math. Tech. Report No.30/79, I.S.I., Calcutta, India.

Das, A.K and Tripathi, T.P. (1981). *A class of sampling strategies for population mean using knowledge on CV of an auxiliary character.* Proceeding of the ISI Golden Jubilee Inter. Confer. On Statistics: Applications and New Directions, Culcatta, 16-19 Dec., 1981, 174-181.

Das, A.K. (1988). *Contribution to the theory of sampling strategies based on auxiliary information.* Ph.D. thesis submitted to B.C.K.V. Mohanpur, Nadia West Bengal, India.

Das, A.K. and Tripathi, T.P. (1978). Use of auxiliary information in estimating the finite population variance. *Sankhya,* Ser. C, 40,139-148.

Das, A.K. and Tripathi, T.P. (1980). Sampling strategies for population mean when the coefficient of variation of an auxiliary character is known. *Sankhya*, Ser. C, 42, 76-86.

Deming, W.E. (1960). *Sampling Design in Business Research.* Wiley Classics Library.

Durbin, J. (1959). A note on the application of Quenouille's method of bias reduction to the estimation of ratios. *Biometrika,* 46,477-480.

Gram, J.P. (1883). Om beregning af en bevoxnings masse ved hjelp af prvetr½r (In Danish). *Tidsskrift Skovbruk*, 6, 137-198.

Graunt, J. (1662), *Natural and Political Observations upon the Bills of Mortality*, London: John Martyn.

Gupta, P. C. (1978). On some quadratic and higher degree ratio and product estimators. *J. Ind. Soc. Agri. Stat.*, *30*, 71-80.

Hald, A. (1998). *A History of Mathematical Statistics from 1750 to 1930*: John Wiley &Sons.

Hanif, M., Ahmad, Z. and Ahmad, M. (2009). Generalized Multivariate Ratio Estimators for Multi-Phase Sampling using Multi-Auxiliary Variables. *Pak. J. Statist.,* 25(4), 615-629.

Hanif, M., Hamad, N. and Shahbaz, M. Q. (2009) A Modified Regression Type Estimator in Survey Sampling, *World App. Sci. J.*, 7(12), 1559-1561.

Hanif, M., Hamad, N. and Shahbaz, M.Q. (2010) Some New Regression Types Estimators in Two Phase Sampling, *World App. Sci. J.*, 8(7), 799-803.

Hanif, M., Haq, I. and Shahbaz, M.Q. (2009) On a New Family of Estimators using Multiple Auxiliary Attributes, *World App. Sci. J.*, 7(11), 1419-1422.

Hanif, M., Haq, I. and Shahbaz, M.Q. (2010) Ratio Estimators using Multiple Auxiliary Attributes, *World App. Sci. J.*, 8(1), 133-136.

Hanif, M., Shahbaz, M.Q. and Ahmad, Z. (2010) Some Improved Estimators in Multiphase sampling, *Pak. J. Statist.*, 26(1), 195-202.

Hansen, M.H and Hurwitz, W.N. and Madow, W.G. (1953). *Sample Survey Methods and Theory.* (Vol. 2), John Wiley and Sons, New York.

Hansen, M.H. and Hurwitz, W.N. (1943). On the Theory of Sampling from Finite Populations. *Ann. Math. Statist.*, 14, 332-362.

Hartley, H.O. and Ross, A. (1954). Unbiased ratio estimators. *Nature,* 174, 270-271.

Hidiroglou, M.A. and Sarndal, C.E. (1995), Use of auxiliary information for two-phase sampling. *Proceedings of the section on Survey Research Methods, American Statistical Association*, Vol. II. 873-878.

Hidiroglou, M.A. and Sarndal, C.E. (1998). Use of auxiliary information for two-phase sampling. *Survey Methodology,* 24(1), 11-20.

Jhajj, H.S., Sharma, M.K. and Grover, L.K. (2006). Dual of ratio estimators of finite population mean obtained on using linear transformation to auxiliary variables. *J. Japan Statist. Soc.*, 36(1), 107-119.

Kaur, P. (1983). Generalized unbiased product estimators. *Pure Appl. Math. Sc.,* 17, 1-2, 67-79.

—. (1984): A combined product estimator in sample surveys. *Biometrical Journal*, 26, 7, 749-753.

Kendall, M.G. and Stuart, A. (1966). *The Advanced Theory of Statistics.* London: Charles Griffin.

Khan, S.U. and Tripathi, T.P. (1967). The use of multi-auxiliary information in double sampling. *J. Ind. Statist. Assoc.*, 5, 42-48.

Khare, B.B. and Srivastava, S.R. (1981). A general regression ratio estimator for the population mean using two auxiliary variables. *Aligarh Journal of Statistics*, 1, 43-51.

Kiregyera, B. (1980). A chain ratio type estimator in finite population double sampling using two auxiliary variables. *Metrika*, 27, 217-223.

Kiregyera, B. (1984). A Regression-Type Estimator using two Auxiliary Variables and Model of Double sampling from Finite Populations. *Metrika*, 31, 215-226.

Kish, L. (1965). *Survey sampling*. John Wiley & Sons, New York.

Kulkarni, S.P. (1978). A note on modified ratio estimator using transformation. *J. Ind. Soc. Agri. Statist.*, 30, 2, 125-128.

Lahiri, D.B. (1951). A method for sample selection providing unbiased ratio estimators. *Bull. Inst. Inter. Statist.,* 33-140.

Laplace, P. S. (1783). Sur les naissances, les mariages et les morts. *Mémoires de l'Académie Royale des Sciences de Paris,* 693–702.

—. (1814a). *Essai philosophique sur les probabilite's*. Paris. Laplace, P.S. (1814b). *Theorie analytique des probabilites.* Paris, 2nd edition.

Lui, K.J. (1990). Modified product estimators of finite population mean infinite sampling. *Commun. in Statist. Theory and Methods,* 19(10), 3700-3807.

Maiti, P. and Tripathi, T.P. (1976). *The use of multivariate auxiliary information in selecting the sample units.* Proceedings of Symposium on recent developments in surveys methodology, I.S.I. Calcutta.

Mickey, M.R. (1959) Some finite population unbiased ratio and regression estimators. *Jour. Amer. Statist. Asso.,* 54. 594-612.

Midzuno, H. (1952). On the sample system with probability proportionate to sum of sizes. *Ann. Inst. of Statist. Math.*, 3, 99-107.

Mohanty, S. (1967). Combination of Regression and Ratio Estimate. *Jour. Ind. Statist. Assoc.*, 5, 16-19.

Mukerjee, R. Rao, T.J. and Vijayan, K. (1987). Regression type estimators using multiple auxiliary information. *Aust. J. Statist.*, 29(3), 244-254.

Murthy, M. N. (1964). Product method of estimation. *Sankhya*, Ser. A. 26, 294-307.

—. (1964). Product method of estimation. *Sankhya*, 26, 294-307.

—. (1967) *Sampling Theory and Methods*. Statistical Publishing Society, Calcutta, India.

Murthy, M.N. and Nanjamma, N.S. (1959). Almost unbiased ratio estimates based on interpenetrating sub-sample estimates. *Sankhya*, 21, 381-392.

Naik, V.D. and Gupta, P.C. (1991) A general class of estimators for estimating population mean using auxiliary information, *Metrika*, 38, 11-17.

Naik, V.D. and Gupta, P.C. (1996). A note on estimation of mean with known population of an auxiliary character. *J. Ind. Soc. Agri. Statist.,* 48(2), 151-158.

Neyman, J. (1934). On the two different aspects of representative method: The method of stratified sampling and the method of purposive selection. *J. Roy. Statist. Soc.*, 97, 558-606.

—. (1938). Contribution to the theory of sampling human populations. *J. Amer. Statist. Assoc.*, 33, 101-116.

—. (1938). *Lecture Notes and Conferences on Mathematical Statistics.* Graduate School, USDA, Washington, DC.

Olkin, I. (1958). Multivariate ratio estimation for finite population. *Biometrika* 45, 154-165.

Pandey, B.N. and Dubey, V. (1988). Modified product estimator using coefficient of variation of auxiliary variable. *Assam Stat. Rev.,* 2, 64-66.

Prasad, B. (1989). Some improved ratio-type estimators of population mean and ratio in finite population sample surveys. *Commun. in Statist., Theo. and Meth.,* 18, 379-392.

Quenouille, M.H. (1956). Note on bias in estimation. *Biometrika,* 43, 353-360.

Raj, D. (1965). On a method of using multi-auxiliary information in sample surveys. *J. Amer. Statist. Assoc.*, 60, 270-277.

Raj, D. (1968). *Sampling Theory.* Tata McGraw Hill.

Rao, J.N.K. (1957). Double ratio estimate in forest survey. *J. Ind. Soc. Agri. Statist.*, 9, 191-204.

—. (1973). On double sampling for stratification and analytic surveys. *Biometrika,* 6, 125-133.

Rao, J.N.K. and Pereira, N.P. (1968). On double ratio estimators. *Sankhya,* Ser. A. 30, 83-90.

Rao, P.S.R.S. (1969). Comparison of four ratio type estimators under a model. *J. Amer. Statist. Assoc.,* 64, 574-580.

Rao, P.S.R.S. and Mudholkar, G.S. (1967). Generalized multivariate estimator for the mean of finite populations. *J. Amer. Statist. Assoc.*, 62, 1009-1012.

Rao, P.S.R.S. and Rao, J.N.K. (1971). Small sample results for ratio estimators. *Biometrika,* 58, 625-630.

Rao, T.J. (1983). *A new class of unbiased product estimators.* Stat Math. Tech. Report No. 15/83, I.S.I., Calcutta India.

Rao, T.J. (1987). On a certain unbiased product estimators. *Commun. Statist. Theo. Meth., 16,* 3631-3640.

Ray, S.K. and Sahai, A. (1980). Efficient families of ratio and product-type estimators. *Biometrika*, 67(1), 211-215.

Ray, S.K., Sahai, A. and Sahai, A. (1979). A note on ratio and product type estimators. *AISM*, 31(1), 141-144.

Ray, S.K. and Singh, R.K. (1981) Difference-cum-ratio type estimators. *J. Ind. Statist. Assoc.*, 19, 147-151.

Reddy, V.N. (1973). On ratio and product method of estimation. *Sankhya, B*, 35, 307-317.

—. (1974). On a transformed ratio method of estimation. *Sankhya, C*, 36, 59-70.

Robson, D.S. (1957). Application of multivariate polykays to the theory of unbiased ratio type estimators. *J. Amer. Statist. Assoc.*, 52, 511-522.

Roy, D.C. (2003). A regression type estimates in two-phase sampling using two auxiliary variables. *Pak. J. Statist.*, 19(3), 281-290.

Royall, R.M. (1970). On Finite Population Sampling Theory under Certain Linear Regression Models. *Biometrika*, 57, 377-387.

Sahai, A. (1979). An efficient variant of the product and ratio estimator. *Statist. Neerlandica,* 33, 27-35.

Sahai, A. and Ray, S.K. (1980). An efficient estimator using auxiliary information. *Metrika,* 27, 27-275.

Sahoo, J. and Sahoo, L.N. (1993). A class of estimators in two-phase sampling using two auxiliary variables. *J. Ind. Soc. Agri. Statist.*, 31, 107-114.

Sahoo, J. and Sahoo, L.N. (1994). On the efficiency of four chain-type estimators in two phase sampling under a model. *Statistics*, 25, 361-366.

Sahoo, J. Sahoo, L.N. and Mohanty, S. (1993). A regression approach to estimation in two phase sampling using two auxiliary variables. *Current Sciences*, 65, 1, 73-75.

Sahoo. J., Sahoo, L.N. and Mohanty, S. (1994). An alternative approach to estimation in two-phase sampling using two auxiliary variables. *Biometrical J.,* 3.293-298.

Samiuddin, M. and Hanif, M. (2006). Estimation in two phase sampling with complete and incomplete information. *Proc. 8th Islamic Countries Conference on Statistical Sciences*. 13, 479-495.

Samiuddin, M. and Hanif, M. (2007). Estimation of population mean in single and two phase sampling with or without additional information. *Pak. J. Statist.*, 23(1), 1-9.

Sarndal, C.E. and Swensson, B. (1987). A general view of estimation for two phases of selection with applications to two-phase sampling and non-response. *International Statistical Review,* 55, 279-294.

Searls, D.T. (1964). The Utilization of a known Coefficient of Variation in the Estimation Procedure. *J. Amer. Statist. Assoc.*, 59, 1225-26.

Sen, A.R. (1952). On the estimate of the variance in sampling with varying probabilities. *J. Ind. Soc. Agri. Statist.,* 5, 119-127.

Shabbir, J. and Gupta, S. (2007). On estimating the finite population mean with known population proportion of an auxiliary variable. *Pak. J. Statist.,* 23(1), 1-9.

Shahbaz, M.Q. and Hanif M. (2009). A general shrinkage estimator in survey sampling. *World Applied Science Journal*, 7(5), 593-596.

Shukla, G.K. (1965). Multivariate regression estimate. *J. Ind. Statist. Assoc.*, 3, 202-211.

—. (1966) An alternative multivariate ratio for finite population. *Cal. Statist. Assoc. Bull.*, 15, 127 -134.

Shukla, N.D. and Pandey, S.K. (*1982*). A note on product estimator, *Pure Appl. Math. Soc.*, 15(12), 97-101.

Singh, D. and Singh, B.D. (1965). Double sampling for stratification on successive occasions. *J. Amer. Statist. Assoc.*, 60, 784-792.

Singh, G.N. and Upadhyaya, L.N. (1995). A class of modified chain type estimators using two auxiliary variables in two phase sampling. *Metron*, 1-III, 117-125.

Singh, H.P. (1987). Class of almost unbiased ratio and product type estimators for finite population mean applying Quenouille's method. *J. Ind. Soc. Agri. Statist.,* 39, 280-288.

—. (1987). On the estimation of population mean when the correlation coefficient is known in two-phase sampling. *Assam Statist. Rev.*, 1(1), 17-21.

Singh, H.P. and Espejo, M.R. (2003). On linear regression and ratio-product estimation of a finite population mean. *The Statistician*, 52, 59-67.

Sin Singh, H.P. and Tailor, R. (2003): Use of known correlation coefficient in estimating the finite population mean. *Statist. in Trans.*, 6(4), 555-560.

Singh, H.P., Upadhyaya, L.N. and Chandra, P. (2004). A General family of estimators for estimating population means using two auxiliary variables in two phase sampling. *Statistics in Transition*, 6, 1055-1077.

Singh, M.P. (1965). On the estimation of ratio and product of population parameters. S*ankhya,* Ser. C, 27, 321-328.

—. (1969). Comparison of Some ratio-cum-product estimators. *Sankhya, Ser. B,* 31, 375-378.

—. (1967). Ratio cum product method of estimation. *Metrika,* 12, 34-43.

—. (1967b). Ratio cum product method of estimation. *Metrika*, 12, 34-42.

Singh, R.K. and Pandey, S.K. (1983). A transformed generalised product-type estimator. *Pure Appl. Math. Sci*., 18(1-2), 67-73.

Sisodia, B.V.S. and Dwivedi, V.K. (1981). A modified ratio estimator using coefficient of variation of auxiliary variable. *Biometrical J.*, 40(6), 753-766.

Sisodia, B.V.S. and Dwivedi, V.K. (1981a). A class of ratio cum product type estimators. *Biometrical J.*, 23, 133-139.

Srivastava, S.K. (1965). An estimate of the mean of a finite population using several auxiliary characters. *J. Ind. Statist. Assoc.*, 3, 189-194.

—. (1966). On ratio and linear regression method of estimation with several auxiliary variables. *J. Ind. Statist. Assoc.,* 4, 66-72.

—. (1967). An estimator using auxiliary information in sample surveys. *Cal. Statist. Assoc. Bull.*, 16, 121-132.

—. (1971). A generalized estimator for the mean of a finite population using multi auxiliary information. *J. Amer. Statist. Assoc.*, 66, 404-407.

—. (1980). A class of estimators using auxiliary information in sample surveys. *Can. J. Statist.,* 8,253-254.

Srivastava, S.K. and Jhajj, H.S. (1980) A class of estimators using auxiliary information for estimating finite population variance. *Sankhya*, C, 42, 87-96.

Srivastava, S.K. and Jhajj, H.S. (1981). A class of estimators of population mean in survey sampling using auxiliary information. *Biometrika,* 68, 341-343.

Srivastava, S.K. and Jhajj, H.S. (1983). A class of estimators of the population means using multi-auxiliary information. *Cal. Statist. Assoc. Bull.,* 32, 47-56.

Srivastava, S.K., Khare, B.B. and Srivastava, S.R. (1990). A generalized chain ratio estimator for mean of finite population. *J. Ind. Soc. Agri. Statist,* 42, 108-117.

Srivenkataraman, T. and Srinath, K.P. (1976). Ratio and product methods of estimation in sample surveys when the two variables are moderately correlated. *Vignana Bharthi.,* 22, 54-58.

Srivenkataramana, T. (1980). A dual to ratio estimator in sample surveys. *Biometrika,* 67, 199-204.

Srivenkataramana, T. and Tracy, D.S. (1979). On ratio and product methods of estimation in sampling. *Statist. Neerlandica*, 33, 37-49.

Srivenkataramana, T. and Tracy, D.S. (1980). An Alternative to ratio method in sample surveys. *Annals of the Institute of Statistical*, 32, 111-120.

Srivenkataramana, T. and Tracy, D.S. (1981). Extending product method of estimation to positive correlation case in surveys. *Austral. J. Statist.,* 23, 95-100.

Sukhatme, P.V. and Sukhatme, B.V., Sukhatme, S. and Ashok, C. (1984). *Sampling Theory of Surveys with Applications*. University Press, Ames, Iowa, U.S.A.

Swain, A.K.P.C. (1970) A note on the use of multiple auxiliary variables in sample surveys. *Trabajos de Estadistica, 21, 135-141.*

Tahir, A. (2009). *An empirical comparison of estimators in two phase sampling.* A thesis submitted to National College of Business Administration and Economics, Lahore.

Tin, M. (1965). Comparison of some ratio estimators. *J. Amer. Statist. Assoc.*, 60, 294-307.

Tiripathi, T.P. (1970). *Contributions to the sampling theory using multivariate information*. Ph.D. thesis submitted to Punjabi University, Patiala, India.

—. (1973). Double sampling for inclusion probabilities and regression method of estimation. *J. Ind. Statist. Assoc.*, 10, 33-46.

—. (1976). On double sampling for multivariate ratio and difference method of estimation. *J. Ind. Statist. Assoc.*, 33, 33-54.

—. (1980). A general class of estimators of population ratio. *Sankhya*, Ser. C. 42, 63-75.

—. (1987). A class of estimators for population mean using multivariate auxiliary information under general sampling designs. *Aligarh J. Statist.,* 7, 49-62.

Tripathi, T.P., Das A.K. and Khare, B.B. (1994).Use of auxiliary information in sample surveys. *A review. Aligarh J. Stat.,* 14, 79-134,

Tripathi, T.P., Singh, H.P. and Upadhyaya, L. N (1988). A generalized method of estimation in double sampling. *J. Ind. Statist. Assoc.,* 26, 91-101.

Upadhyaya, L.N. and Singh, H.P. (1984). On families of ratio and product-type estimators. *J. Ind. Soc. Statist. and Oper. Res.*, 5, (1-4), 57-61.

Upadhyaya, L.N. and Singh, G.N. (2001), Chain type estimators using transformed auxiliary variable in two-phase sampling. *Advances in Modeling and Analysis*, 38(1-2), 1-10.

Vos, J.W.E. (1980). Mixing of direct ratio and product method estimators. *Statist. Neerlandica,* 34, 209-218.

Walsh, J.E. (1970). Generalization of ratio estimate for population total. *Sankhya*, (A), 32, 99-106.

Watson, D.J. (1937). The estimation of lead areas. *J. Agri. Sci.*, 27, 475.

Williams, W.H. (1963). The precision of some unbiased regression estimators. *Biometrics*, 19(2), 352-361.

Yates, F. (1960). *Sampling Methods for Censuses and Surveys*, 2nd Ed. (London: Charles Griffin).

APPENDICES

Table A.1:
Ranked MSE of Ratio and Regression Estimator in Two-phase Sampling (All information from population)

Est./ POP	1	2	3	4	5	6	7	8	9	10	11	12	13	14	15	16	Avg.	Rank
$t_{1(2)}$	2	3	2	2	2	3	5	2	2	2	7	2	7	4	2	2	3.06	2
$t_{2(2)}$	5	5	5	5	5	5	4	5	4	4	5	4	5	4	5	4	4.63	5
$t_{3(2)}$	4	4	3	5	4	4	2	4	5	5	4	5	1	7	3	5	4.06	3
$t_{4(2)}$	7	7	7	7	7	7	6	7	3	3	1	3	3	2	4	3	4.81	6
$t_{5(2)}$	3	2	4	3	3	1	7	3	7	7	3	7	4	6	7	7	4.63	5
$t_{6(2)}$	6	6	6	6	6	6	3	6	6	6	2	7	2	1	6	6	5.06	7
$t_{7(2)}$	1	1	1	2	1	2	1	1	2	2	7	2	7	4	2	2	2.38	1

Table A.2:
Ranked MSE of Ratio and Regression Estimator in Two-phase Sampling (Information from first and second phase)

Est./POP	3	4	5	7	11	12	13	14	15	16	Avg.	Rank
$t_{1(2)}$	2	3	2	7	7	2	6	6	2	2	3.9	4
$t_{2(2)}$	4	5	5	6	7	4	7	5	5	4	5.2	6
$t_{3(2)}$	5	1	3	1	3	5	1	7	3	5	3.4	2
$t_{4(2)}$	7	7	7	3	3	7	5	4	7	7	5.7	7
$t_{5(2)}$	3	4	4	5	3	3	3	3	4	3	3.5	3
$t_{6(2)}$	6	6	6	2	3	6	4	1	6	6	4.6	5
$t_{7(2)}$	1	2	1	4	3	1	2	3	1	1	1.9	1

**Table A.3:
Ranked MSE of Ratio and Regression Estimator in Two-phase Sampling (All information from second phase)**

Est./POP	3	4	5	7	11	12	13	14	15	16	Avg.	Rank
$t_{1(2)}$	2	4	2	7	3	2	6	7	2	2	3.7	4
$t_{2(2)}$	4	5	5	6	4	4	7	6	5	4	5.0	6
$t_{3(2)}$	5	2	3	1	5	5	1	2	3	5	3.2	2
$t_{4(2)}$	7	7	7	3	7	7	5	3	7	7	6.0	7
$t_{5(2)}$	3	3	4	5	2	3	3	5	4	3	3.5	3
$t_{6(2)}$	6	6	7	2	6	6	4	1	6	6	5.0	6
$t_{7(2)}$	1	1	1	4	1	1	2	4	1	1	1.7	1

**Table A.4:
Ranked MSE of Mohanty Estimator in Two-phase Sampling (All information from population)**

Est./ Pop.	1	2	3	4	5	6	7	8	9	10	11	12	13	14	15	16	Avg.	Rank
$t_{8(2)}$	3	3	3	3	4	3	7	3	4	4	4	4	2	5	4	3	3.69	3
$t_{9(2)}$	6	5	5	6	6	6	5	6	5	5	3	5	3	3	6	5	5.00	6
$t_{10(2)}$	5	6	6	4	3	5	1	5	2	3	7	2	4	1	3	4	3.81	4
$t_{11(2)}$	7	7	7	7	7	7	3	7	8	8	6	8	7	7	7	7	6.88	7
$t_{12(2)}$	2	2	2	2	2	1	6	2	6	6	2	6	5	6	2	6	3.63	2
$t_{13(2)}$	8	8	8	8	8	8	4	8	7	7	8	7	8	8	8	8	7.56	8
$t_{14(2)}$	4	4	4	5	5	4	8	4	3	2	5	3	1	4	5	2	3.94	5
$t_{15(2)}$	1	1	1	1	1	2	2	1	1	1	1	1	6	2	1	1	1.50	1

Table A.5:
Ranked MSE of Mohanty Estimator in Two-phase Sampling
(Information at both phases)

Est./Pop.	3	4	5	7	11	12	13	14	15	16	Avg.	Rank
$t_{8(2)}$	3	3	3	2	3	4	2	5	3	4	3.2	3
$t_{9(2)}$	5	5	6	5	4	6	5	4	6	5	5.1	6
$t_{10(2)}$	4	7	5	4	7	2	4	1	4	3	4.1	5
$t_{11(2)}$	7	4	7	7	6	8	8	7	7	7	6.8	7
$t_{12(2)}$	6	1	2	1	5	3	3	3	2	6	3.2	3
$t_{13(2)}$	8	8	8	8	8	7	7	8	8	8	7.8	8
$t_{14(2)}$	2	6	4	3	2	5	1	6	5	2	3.6	4
$t_{15(2)}$	1	2	1	6	1	1	6	2	1	1	2.2	1

Table A.6:
Ranked MSE of Mohanty Estimator in Two-phase Sampling
(Information from second phase only)

Est./Pop.	3	4	5	7	11	12	13	14	15	16	Avg.	Rank
$t_{8(2)}$	4	4	3	2	2	4	2	4	3	4	3.2	3
$t_{9(2)}$	5	5	7	5	4	5	5	5	6	5	5.2	6
$t_{10(2)}$	2	6	4	4	5	2	4	1	5	2	3.5	4
$t_{11(2)}$	7	3	5	7	7	8	8	7	7	7	6.6	7
$t_{12(2)}$	6	2	2	1	6	3	1	2	2	6	3.1	2
$t_{13(2)}$	8	8	8	8	8	7	7	8	8	8	7.8	8
$t_{14(2)}$	3	7	6	3	1	6	3	6	4	3	4.2	5
$t_{15(2)}$	1	1	1	6	3	1	6	3	1	1	2.4	1

Table A.7:
Ranked MSE of Chain Estimators in Two-phase Sampling
(All information from population)

Est./ Pop.	1	2	3	4	5	6	7	8	9	10	11	12	13	14	15	16	Avg.	Rank
$t_{16(2)}$	1	2	1	1	1	3	2	1	2	3	4	3	8	3	1	2	2.38	1
$t_{17(2)}$	2	1	2	2	2	1	9	2	5	6	1	5	4	5	2	5	3.38	2
$t_{18(2)}$	12	12	13	12	12	12	3	12	13	13	13	13	10	11	12	13	11.63	12
$t_{19(2)}$	11	11	12	11	11	11	4	11	11	12	12	11	5	4	11	12	10.00	11
$t_{20(2)}$	6	5	5	5	6	7	1	6	4	4	7	6	11	7	5	4	5.56	5
$t_{21(2)}$	10	10	9	10	10	10	8	10	10	9	8	10	12	12	10	9	9.81	10
$t_{22(2)}$	7	6	8	6	7	6	10	7	8	8	5	8	2	6	6	8	6.75	8
$t_{23(2)}$	13	13	11	13	13	13	11	13	12	10	11	12	13	13	13	11	12.19	13
$t_{24(2)}$	9	9	10	9	9	9	7	9	9	11	10	9	1	10	9	10	8.75	9
$t_{25(2)}$	5	8	6	7	5	5	13	5	3	1	9	1	6	9	7	3	5.81	6
$t_{26(2)}$	4	3	4	4	4	2	12	4	6	5	3	5	3	8	4	6	4.81	4
$t_{27(2)}$	3	4	3	3	3	4	6	3	1	2	7	2	9	3	3	1	3.56	3
$t_{28(2)}$	8	7	7	8	8	8	5	8	7	7	3	7	7	1	8	7	6.63	7

**Table A.8:
Ranked MSE of Chain Estimator in Two-phase Sampling
(Information at both phases)**

Est./ Pop.	3	4	5	7	11	12	13	14	15	16	Avg.	Rank
$t_{16(2)}$	2	2	1	8	4	2	7	2	1	3	3.2	2
$t_{17(2)}$	4	1	2	4	1	4	3	4	2	5	3.0	1
$t_{18(2)}$	13	12	12	10	13	13	11	11	12	13	12.0	12
$t_{19(2)}$	12	11	11	6	12	11	9	7	11	12	10.2	10
$t_{20(2)}$	6	5	6	11	7	3	10	10	5	4	6.7	8
$t_{21(2)}$	9	10	10	12	9	10	12	12	10	9	10.3	11
$t_{22(2)}$	8	6	7	3	6	8	4	5	7	8	6.2	6
$t_{23(2)}$	11	13	13	13	10	12	13	13	13	11	12.2	13
$t_{24(2)}$	10	8	9	5	11	9	5	1	9	10	7.7	9
$t_{25(2)}$	3	9	5	2	8	5	1	9	6	2	5.0	5
$t_{26(2)}$	5	3	4	1	3	6	2	8	4	6	4.2	4
$t_{27(2)}$	1	4	3	9	5	1	6	6	3	1	3.9	3
$t_{28(2)}$	7	7	8	7	3	7	8	3	8	7	6.5	7

Table A.9:
Ranked MSE of Chain Estimator in Two-phase Sampling
(Information from second phase only)

Est./Pop.	3	4	5	7	11	12	13	14	15	16	Avg.	Rank
$t_{16(2)}$	3	2	1	8	2	1	7	5	1	2	3.2	2
$t_{17(2)}$	5	1	2	4	4	4	3	2	2	5	3.2	2
$t_{18(2)}$	13	12	12	10	13	12	11	11	12	13	11.9	12
$t_{19(2)}$	11	11	11	6	12	11	9	8	11	12	10.2	10
$t_{20(2)}$	6	6	5	11	6	3	10	10	6	4	6.7	7
$t_{21(2)}$	10	10	10	12	9	10	12	12	10	9	10.4	11
$t_{22(2)}$	8	5	6	3	8	8	4	3	7	8	6.0	6
$t_{23(2)}$	12	13	13	13	11	13	13	13	13	11	12.5	13
$t_{24(2)}$	9	7	9	5	10	9	5	1	9	10	7.4	9
$t_{25(2)}$	1	9	8	2	3	6	1	9	5	3	4.7	5
$t_{26(2)}$	4	3	4	1	5	5	2	4	4	6	3.8	3
$t_{27(2)}$	2	4	3	9	2	2	6	7	3	1	3.9	4
$t_{28(2)}$	7	8	7	7	7	7	8	6	8	7	7.2	8

Table A.10:
Ranked MSE of Kiregyera Estimators in Two-phase Sampling
(All information from population)

Est./ Pop.	1	2	3	4	5	6	7	8	9	10	11	12	13	14	15	16	Avg.	Rank
$t_{29(2)}$	2	1	2	3	1	3	3	2	2	2	1	1	6	2	1	2	2.13	2
$t_{30(2)}$	4	2	4	4	4	1	2	4	5	5	3	5	1	1	4	4	3.31	3
$t_{31(2)}$	3	5	3	2	3	5	5	3	3	3	5	3	4	4	3	3	3.56	4
$t_{32(2)}$	5	4	5	5	5	2	7	5	6	6	4	6	2	6	5	6	4.94	5
$t_{33(2)}$	1	3	1	1	2	4	1	1	1	1	2	2	3	3	2	1	1.81	1
$t_{34(2)}$	6	6	6	6	6	6	4	6	4	4	6	4	5	5	6	5	5.31	6
$t_{35(2)}$	7	7	7	7	7	7	6	7	7	7	7	7	7	7	7	7	6.94	7

Table A.11:
Ranked MSE of Kiregyera Estimators in Two-phase Sampling
(Information at both phases)

Est./Pop.	3	4	5	7	11	12	13	14	15	16	Avg.	Rank
$t_{29(2)}$	1	4	2	7	1	2	6	2	2	1	2.80	2
$t_{30(2)}$	4	2	3	2	3	5	2	1	4	5	3.10	3
$t_{31(2)}$	3	5	4	6	5	3	4	4	3	3	4.00	5
$t_{32(2)}$	5	1	5	1	4	6	1	6	5	6	4.00	5
$t_{33(2)}$	2	3	1	4	2	1	3	3	1	2	2.20	1
$t_{34(2)}$	6	6	6	5	6	4	5	5	6	4	5.30	6
$t_{35(2)}$	7	7	7	3	7	7	7	7	7	7	6.60	7

Table A.12:
Ranked MSE of Kiregyera Estimators in Two-phase Sampling (Information from second phase only)

Est./Pop.	3	4	5	7	11	12	13	14	15	16	Avg.	Rank
$t_{29(2)}$	1	3	2	6	1	3	6	2	2	1	2.7	2
$t_{30(2)}$	4	1	5	2	3	5	2	1	4	5	3.2	3
$t_{31(2)}$	3	5	3	5	4	2	4	5	3	3	3.7	4
$t_{32(2)}$	6	4	4	1	5	6	1	3	5	6	4.1	5
$t_{33(2)}$	2	2	1	3	2	1	3	4	1	2	2.1	1
$t_{34(2)}$	5	6	6	4	6	4	5	6	6	4	5.2	6
$t_{35(2)}$	7	7	7	7	7	7	7	7	7	7	7.0	7

Table A.13:
Ranked MSE of Sahoo, Sahoo and Mohanty Estimators in Two-phase Sampling (All information from population)

Est./ Pop.	1	2	3	4	5	6	7	8	9	10	11	12	13	14	15	16	Avg.	Rank
$t_{39(2)}$	2	2	2	2	2	3	3	2	2	2	3	2	3	3	2	2	2.31	2
$t_{40(2)}$	3	3	4	3	3	1	6	3	5	6	4	5	2	2	3	4	3.56	5
$t_{41(2)}$	5	5	5	5	5	5	2	5	4	3	2	3	1	1	5	5	3.81	4
$t_{42(2)}$	7	7	7	7	7	7	4	7	7	7	7	7	5	6	7	7	6.63	7
$t_{43(2)}$	6	6	6	6	6	6	1	6	6	4	6	6	4	6	6	6	5.44	6
$t_{44(2)}$	4	4	1	4	4	4	5	4	1	1	1	1	6	4	4	1	3.06	1
$t_{45(2)}$	1	1	3	1	1	2	7	1	3	5	5	4	7	7	1	3	3.25	3

Table A.14:
Ranked MSE of Sahoo, Sahoo and Mohanty Estimators in Two-phase Sampling (Information at both phases)

Est./Pop.	3	4	5	7	11	12	13	14	15	16	Avg.	Rank
$t_{39(2)}$	2	3	2	3	3	2	3	3	2	2	2.50	1
$t_{40(2)}$	5	1	3	2	4	6	2	2	3	5	3.30	4
$t_{41(2)}$	4	5	5	1	1	5	1	1	5	4	3.20	3
$t_{42(2)}$	7	7	7	7	7	7	7	7	7	7	7.00	7
$t_{43(2)}$	6	6	6	6	5	4	6	6	6	6	5.70	6
$t_{44(2)}$	1	4	4	4	2	1	4	4	4	1	2.90	2
$t_{45(2)}$	3	2	1	5	6	3	5	5	1	3	3.40	5

Table A.15:
Ranked MSE of Sahoo, Sahoo and Mohanty Estimators in Two-phase Sampling (Information from second phase only)

Est./Pop.	3	4	5	7	11	12	13	14	15	16	Avg.	Rank
$t_{39(2)}$	2	3	2	3	3	2	3	3	2	2	2.50	1
$t_{40(2)}$	5	1	3	2	4	6	2	2	3	5	3.30	4
$t_{41(2)}$	4	5	5	1	1	5	1	1	5	4	3.20	3
$t_{42(2)}$	7	7	7	7	7	7	7	7	7	7	7.00	7
$t_{43(2)}$	6	6	6	6	5	4	6	6	6	6	5.70	6
$t_{44(2)}$	1	4	4	4	2	1	4	4	4	1	2.90	2
$t_{45(2)}$	3	2	1	5	6	3	5	5	1	3	3.40	5

Table A.16:
Ranked MSE of Upadhyaya and Singh Estimators in Two-phase Sampling (All information from population)

Est./Pop.	3	4	5	7	11	12	13	14	15	16	Avg.	Rank
$t_{46(2)}$	1	1	1	4	3	1	5	1	1	1	1.9	1
$t_{47(2)}$	4	3	3	7	2	5	4	6	3	5	4.2	4
$t_{48(2)}$	2	2	2	3	4	2	6	2	2	2	2.7	2
$t_{49(2)}$	3	4	4	5	1	6	3	5	4	6	4.1	3
$t_{50(2)}$	6	6	6	2	7	4	7	3	6	4	5.1	6
$t_{51(2)}$	8	8	8	8	6	8	1	8	8	8	7.1	8
$t_{52(2)}$	5	5	5	1	8	3	8	4	5	3	4.7	5
$t_{53(2)}$	7	7	7	6	5	7	2	7	7	7	6.2	7

Table A.17:
Ranked MSE of Upadhyaya and Singh Estimators in Two-phase Sampling (Information at both phases)

Est./Pop.	3	4	5	7	11	12	13	14	15	16	Avg.	Rank
$t_{46(2)}$	1	4	1	5	3	1	6	2	1	1	2.5	1
$t_{47(2)}$	4	2	3	4	2	6	2	4	3	5	3.5	4
$t_{48(2)}$	2	3	2	6	4	2	5	1	2	2	2.9	2
$t_{49(2)}$	3	1	4	3	1	5	1	3	4	6	3.1	3
$t_{50(2)}$	6	6	6	8	7	3	7	8	6	4	6.1	7
$t_{51(2)}$	8	8	8	1	6	7	3	6	8	8	6.3	8
$t_{52(2)}$	5	5	5	7	8	4	8	7	5	3	5.7	5
$t_{53(2)}$	7	7	7	2	5	8	4	5	7	7	5.9	6

Table A.18:
Ranked MSE of Upadhyaya and Singh Estimators in Two-phase Sampling (Information from second phase only)

Est./Pop.	3	4	5	7	11	12	13	14	15	16	Avg.	Rank
$t_{46(2)}$	2	3	2	5	1	6	5	6	2	2	3.4	4
$t_{47(2)}$	4	2	3	4	3	5	1	3	3	5	3.3	3
$t_{48(2)}$	1	4	1	6	2	1	6	7	1	1	3	1
$t_{49(2)}$	3	1	4	3	4	4	2	1	4	6	3.2	2
$t_{50(2)}$	5	7	5	8	5	2	7	8	5	3	5.5	5
$t_{51(2)}$	7	6	7	2	7	7	4	4	8	8	6	8
$t_{52(2)}$	6	8	6	7	6	3	8	5	6	4	5.9	7
$t_{53(2)}$	8	5	8	1	8	8	3	2	7	7	5.7	6